WARM-UP EXERCISE

This warm-up exercise does not require any written responses, but take a few minutes to consider the following questions about the purpose and importance of models.

How badly would our ability to think through problems be hampered if you were not able to rely on models such as diagrams, mathematical formulas and symbols, and maps?

How well would we be able to communicate without the aid of such models as sketches, symbols, diagrams, and so on?

How well would we be able to predict the performance characteristics of a new airplane without the aid of scale models and wind tunnels?

How well could the construction of a complex building be controlled without the model of the building as it exists in the plans and engineering drawings?

How sure could we be that our astronauts would be well trained for space flights without the models of the spacecraft used to simulate actual flight conditions?

Notes: ______________________________

Learning Segment 7
MODELS: AN INTRODUCTION

LEARNING OBJECTIVES

- Understand the meaning of the terms *iconic representation, diagrammatic representation, graphic representation, mathematical representation, iconic simulation, analog simulation, digital simulation, participative simulation, model,* and *gaming.*

- Apply digital simulation to simple problems.

PHYSICAL REPRESENTATIONS

The following objects have a characteristic in common: toy train, global replica of the earth, statue, molecular model, and toy airplane. Each is three-dimensional *representation* of a physical reality. There are two-dimensional equivalents, as exemplified by photographs, sketches, and blueprints. Since these two- and three-dimensional representations bear a physical resemblance to their real-life counterparts, they are referred to as *iconic representations*. Engineers make frequent use of iconic representations; you can see why the one pictured in Figure 1 is a valuable aid. Imagine trying to visualize what this structure looks like from 189 pages of complicated engineering drawings!

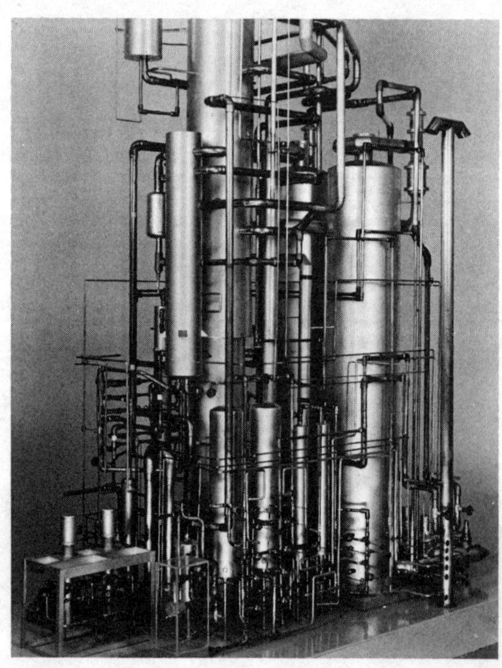

Figure 1. An iconic representation of an oxygen-production facility. This table-top model aids the designers in laying out piping so that areas will be accessible for maintenance and repair. It is also used as a visualization aid to them and to persons to whom they must explain the design.

MODERN ENGINEERING
More Engineering Skills

Edward V. Krick

Lafayette College
Easton, Pennsylvania

Wiley Professional Development Programs

Advisory Editor
Steven C. Wheelwright
Harvard Business School

John Wiley & Sons, Inc.
New York • London • Sydney • Toronto

Library of Congress Catalogue Card Number: 76-10038

ISBN 0-471-01702-7

Printed in the United States of America.

10 9 8 7 6 5 4 3 2 1

TECHNICAL TERMS

Write a brief definition for the following term.

Iconic representation

____

____

____

____

____

____

____

____

An iconic representation is a model that looks like the real-life thing that is being described by the model. An iconic representation may appear in three dimensions (such as a model train or a globe) or in two dimensions (such as a photograph or a map).

The diagram for a football play, the schematic diagram for an electrical circuit, and Figure 2 are *diagrammatic representations*. In each instance, a configuration of lines and symbols represents the structure or behavior of a real-life counterpart.

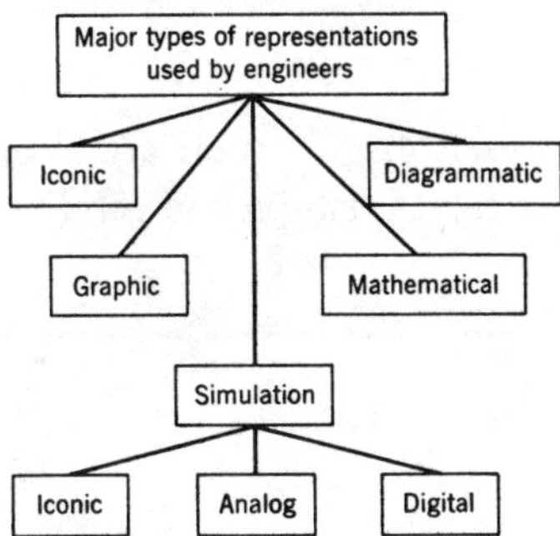

Figure 2. A diagrammatic representation of the contents of this learning segment.

A diagram such as Figure 3 is certainly helpful to those planning and managing the project, especially because of the number of interrelationships. As you can imagine, with engineering devices, structures, and processes becoming so complex, diagrams must be relied on extensively in designing these systems and in communicating their make-up to others.

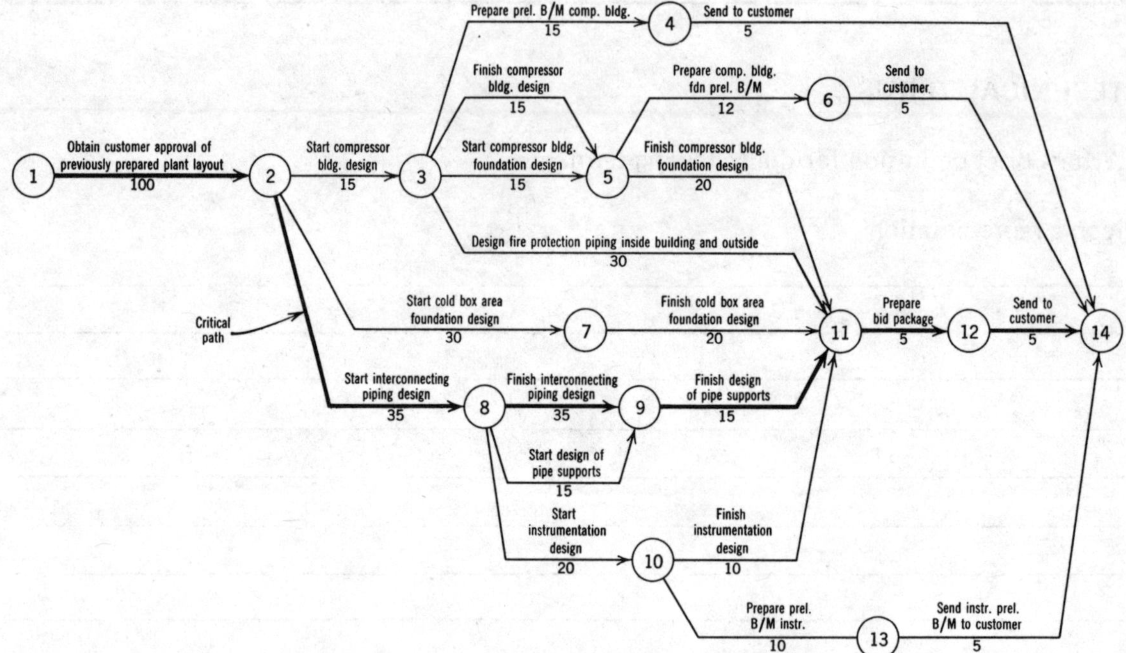

Figure 3. A diagrammatic representation of a portion of an engineering project. Arrows leading to a circle represent those phases of the project that must be completed before activities represented by arrows leading away from that circle can begin. The number beneath an arrow is the number of days that phase of the project is expected to require for completion. There is one "path" through this network, referred to as the critical path, which determines the minimum period of time in which the total project can be completed. This type of representation, commonly referred to as a critical path diagram, is a very useful aid in planning and controlling large scale design and construction projects. The project diagrammed above involves design of the foundation, piping system, and building for a chemical plant. As critical path diagrams go, this is simple. But even this one demonstrates the value of diagrammatic representations as aids to visualization of relationships.

Then there is the familiar *graphic representation*. You are already familiar with the usefulness of graphs in aiding you to visualize relationships and relative magnitudes.

TECHNICAL TERMS

Write a brief definition for each term.

1. Diagrammatic representation

____

____

____

2. Graphic representation

---

1. A diagrammatic representation is model, consisting of a set of lines and symbols, that conveys the structure or behavior of some real-life counterpart.

2. A graphic representation is a model, consisting of lines, numbers, and other symbols, that conveys the relationships and relative magnitudes of the components of some real-life counterpart.

MATHEMATICAL REPRESENTATIONS

The mathematical expression shown here is a representation.

$$V = \frac{mkT}{p}$$

The letter m represents the mass of a quantity of gas, T represent its temperature, p represents the pressure applied, V represents the volume occupied by that gas, and k is a constant. Together these letters represent what happens to one of the properties when a specified change occurs in another. This *mathematical representation* provides a means of predicting one property, given specific values of the others. Through thepower of mathematics, predictions can be made of many natural phenomena as well as of the behavior of man-made mechanisms and structures. By employing the conventions of mathematics and by assigning symbols to represent relevant properties of the real thing, mathematical expressions can be developed for making useful predictions of what to expect under given conditions.

The study of mathematics will provide you with a repertory of ready-made mathematical representations (parabolic function, sine function, etc.). It also equips you to derive special mathematical expressions to fit situations that cannot be represented satisfactorily by ready-made mathematical functions. This is a very important skill for engineers.

Mathematics provides a powerful means of representation, serving as an effective means of *prediction* and as a concise, commonly understood language for *communication*. Its conventions also make it an extremely useful *medium for reasoning*. Can you imagine trying to perform in words some of the logic and the manipulations that you can do so conveniently through the symbolism of mathematics? In addition, training in mathematics strengthens your ability to think clearly and logically. In view of the utility of mathematics as a means of prediction, communication, and reasoning, the heavy emphasis given to this subject in engineering education is readily understandable.

LET'S BE SURE ABOUT THIS

Mathematical representation is an extremely useful tool for engineers. List the three uses (stressed in the preceeding paragraph) to which we put mathematics.

1. ______________________________

2. ______________________________

3. ______________________________

1. Prediction 2. Communication 3. Reasoning

SIMULATION

Iconic Simulations

We have seen how an iconic representation can be used to predict the behavior of its real counterpart. A model of a proposed aircraft is subjected to high-velocity winds in the wind tunnel in order to predict how a real plane of that design will perform in actual flight. What the wind tunnel and plane model are to the aircraft designer, the facilities pictured in Figures 4 and 5 to the designers of bridges, tall buildings, power plants, and ocean-going vessels.

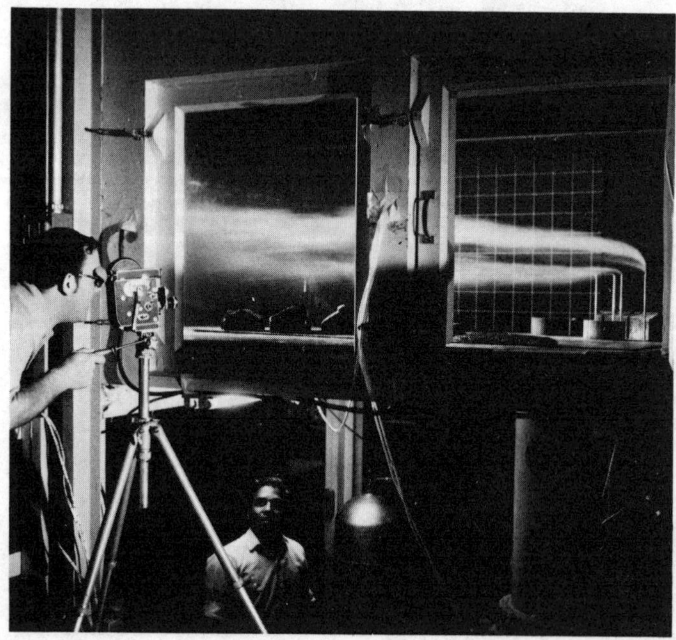

Figure 4. Wind tunnels are useful for more than testing aircraft. Here, through the viewing window of a wind tunnel, you see a scale model of a power plant under design. Engineers are experimenting with different stack heights to determine what is necessary to disperse the stack emissions satisfactorily. Also tested in wind tunnels are bridge designs for wind-induced oscillations, models of tall buildings to predict the wind currents they will create, and models of ocean liners for optimum location of stacks.

Figure 5. In this basin, models of ocean-going vessels are tested for maneuverability and seaworthiness. The carriage that tows and maneuvers the models rides along a 115-meter bridge that itself is movable. Special machines generate waves of specified size and frequency. This simulation setup enables engineers to predict the full-scale performance of proposed ship designs on the high seas.

The same is true for engineers planning flood control measures (Figure 6). The process of *experimenting* with a *representation* of the real thing is called *simulation*. When the experiments are performed on iconic representations, the process is called *iconic simulation*.

Figure 6. This is one section of an enormous working scale model–the largest in existance, representing the Mississippi River and its tributaries from Sioux City, Iowa, to the Gulf of Mexico. This simulator is used to predict the local and system-wide effects of proposed dams, diversionary channels, and other construction projects.

TECHNICAL TERMS

Write a brief definition for each term.

1. Simulation

____

____

2. Iconic simulation

____

____

1. Simulation, generally, is the process of experimenting with any representation of a real thing.

2. Iconic simulation is the process of experimenting with an iconic representation, that is, a representation that looks similar to some real-world counterpart.

Analog Simulations

There are other forms of simulation in which the experiments are performed on representations that bear a behavioral but not a physical resemblance to their real-life counterparts. One of these is called *analog simulation*, another is called *digital simulation*.

An example of analog simulation is the electronic device used by an engineer who is designing a traffic control system. Special electrical circuits represent the city's traffic arteries, while electrical pulses represent vehicles. With this simulator the engineer experiments with different traffic control schemes. Here electrical pulses behave analogously to autos moving about the city, even though pulses and wires in no way physically resemble autos and streets. In the analog simulator pictured in Figure 7, water behaves analogously to air. It

Figure 7. An analog simulator. In this instance, water behaves analogously to air and serves as the medium for experimentation. The device simulates air flow through the diffuser blades (represented by wedge-shaped objects) in the compressor stage of a gas turbine under development. Water containing a dye that makes the path of flow readily observable is diffused from the center at one thousandth of the velocity of the gas it represents. By experimenting with different wedge shapes, angles, and locations, the investigators will learn how to maximize effectiveness.

enables designers of gas turbines to test their ideas quickly and cheaply. Thus, in analog simulation, a medium that behaves analogously to the real phenomenon is employed as a vehicle for experimentation. Electricity is the medium often used.

TECHNICAL TERMS

Write a brief definition for the following term.

Analog simulation

An analog simulation is an experiment with a medium that behaves in a manner that resembles the behavior of the real-world counterpart of the analog medium. The analog may not physically resemble the real-world counterpart in appearance (as is the case with iconic simulation).

Digital Simulations

Digital simulations are best introduced by an example. A university plagued by auto parking problems has engaged a consulting engineer to improve the situation. One of his proposals is to separate the drivers who consistently use their parking spaces the full working day from those who use their spaces sporadically (e.g., a few hours one morning, all the next afternoon, an hour the following morning). The full-day users will be assigned regular *spaces*, but the sporadic users will be grouped and assigned to a *lot* where they will use any available space. The engineer theorizes that more of the sporadic drivers can be assigned to a lot than there are spaces, with negligible risk of the lot overflowing. Notice, however, that this is an untested idea.

THINK IT THROUGH

How would the consulting engineer verify his hypothesis?

Answers to this exercise appear in the following paragraphs.

One way the engineer could verify his hypothesis would be to set up a special lot, assign only sporadic users to it, and then observe it for a significant period of time to see what happens. This is too costly, too involved, and too time-consuming. A logical alternative to this real-life experiment to test the theory by simulation, which the engineer did. In this simulation he assumed a 20-space lot to which 25 drivers were assigned. His procedure and results are described in Figures 8 through 11.

Figure 8. This sequence of steps diagrammed and explained below is repeated 25 times to complete the simulation of one day of parking-lot operation. Then another day is simulated, and another, until sufficient experience in the operation of this hypothetical parking lot has been synthesized. The spinner device is used to recreate the element of chance associated with events. The use of this device and the drawing-from-the-bowl procedure are ways of making random selections from a collection of numbers, appropriately referred to as Monte Carlo Procedures. The results for one simulated day are shown in Figure 9. The method of recording and tabulating the numbers for this digital simulation is illustrated in Figure 10.

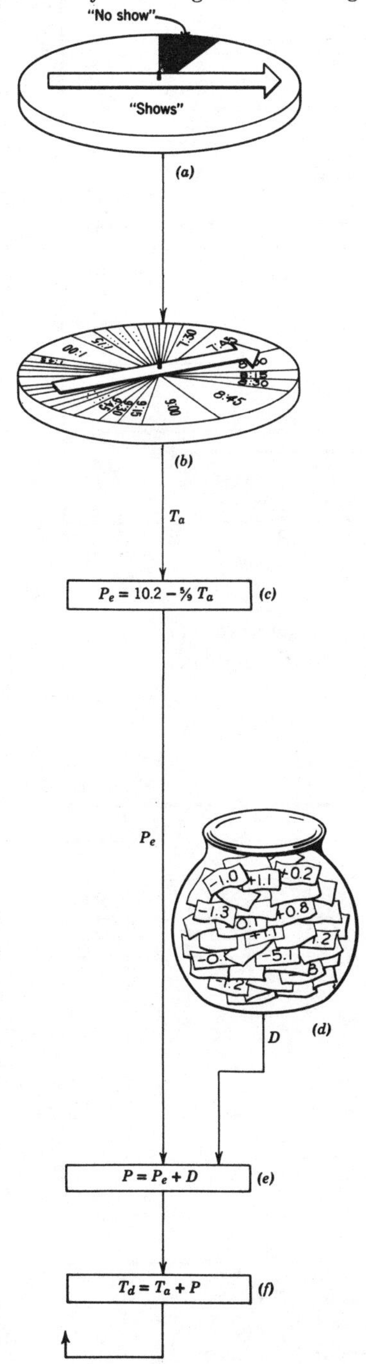

(*a*) From a survey the engineer estimates the probability that a driver will not use the lot on any given day as 8 chances in 100. This "no show" probability of 0.08 is taken into account by using this roulette-type device. Beneath the pointer is a pie chart having a "no show" sector of 28.8 degrees (0.08 × 360 degrees). This pointer is spun for each of the 25 drivers, to determine whether he uses the lot on that day.

(*b*) The likelihood of a driver arriving is not the same throughout the day. To determine how arrival frequency changes with time of day some actual observations were made. These data were converted to this pie chart in which each sector is proportional to the average arrival frequency of drivers for the indicated time of day. This pointer is spun to select an arrival time (T_a) for each driver.

(*c*) The length of time an auto remains on the lot depends on the time it arrives. To learn the nature of this relationship some "park times" were measured resulting in the graph below. A straight line was fitted to this data. From the

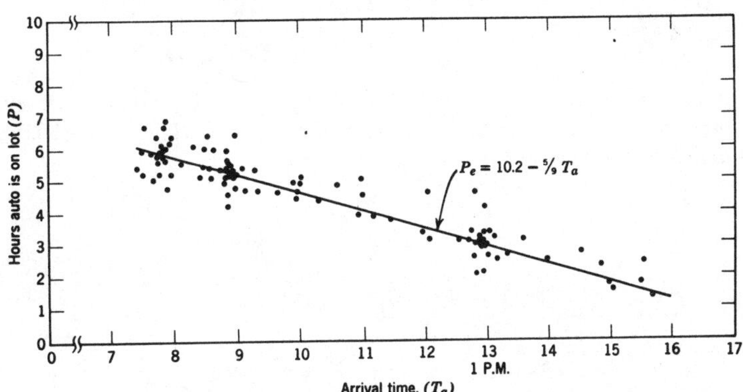

equation of this line the *expected* (most likely) park time (P_e) can be computed, given the driver's time of arrival. (P_e is the average parking time for many drivers arriving at a given time of day.)

(*d*) From the above graph it is apparent that the *actual* periods of time autos remain on the lot vary considerably from the *expected* times (i.e., from the straight line). The vertical deviation of each point from the straight line was measured (plus if above the line, minus if below it). Each of these deviation values was recorded on a separate slip of paper and placed in a bowl. In the simulation a slip is selected randomly, the *D* value noted, and the slip returned to the bowl.

(*e*) Then this *D* value is added to the *expected* park time (P_e) to obtain the *actual* park time (P).

(*f*) The departure time (T_d) of a given driver is calculated by adding his park time to his time of arrival.

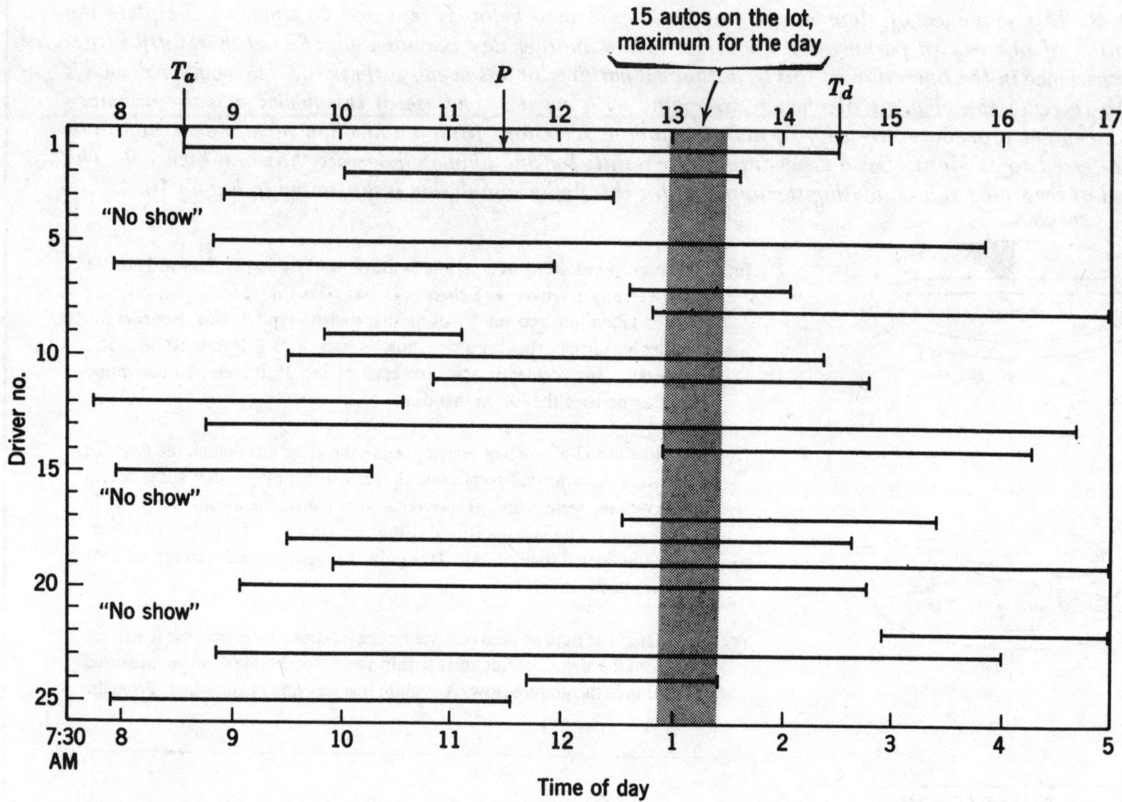

Figure 9. A graphic representation of one day of simulated parking-lot operation. The maximum number of autos is determined by noting the maximum number of simultaneously overlapping bars.

TABULATION SHEET FOR PARKING SIMULATION													Day No. 1
↓ Identification of Row	Source of Table Entries	Driver No.											
		1	2	3	4	5	6	7	21	22	23	24	25
Show or no show?	Spinner												
T_a (arrival time)	Spinner	8.5	10.0	8.9									
P_e (expected park time)	$P_e = 10.2 - .56 T_a$ (or use graph)	5.4	4.6										
D (chance deviation)	Chip drawn from bowl	+.5	-1.1										
P (actual park time)	$P = P_e + D$	5.9	3.5										
T_d (departure time)	$T_d = T_a + P$	14.4 (2:24)	13.5 (1:30)										

Figure 10. Here is the crux of the digital simulation. In executing this simulation, the operator performs the indicated numerical operations, using this sheet as a running record of the experience he has synthesized. (Hours are in decimal form, so that 12 midnight = 0, 9:30 a.m. = 9.5, and 3:00 p.m. = 15.0.)

Figure 11. This graph is a summary of 100 daily maximums. Each daily maximum is determined as explained in Figures 9 and 10. Note that on only one of the 100 days simulated did the lot overflow, even though 25 drivers were assigned to a 20-space lot. Furthermore, on only 4 percent of the days did the daily maximums exceed 19 autos, and on only 8 percent of the days would the lot have overflowed if there had been 18 spaces. From this it is apparent that more drivers of the sporadic type can be assigned than there are spaces on the lot, resulting in better utilization with negligible inconvenience to the users. Furthermore, the client can decide what probability of overflow he will tolerate; then the engineer can use this graph to determine how many drivers can be assigned to a given lot without exceeding this probability. Suppose university officials say that they don't mind if the lot overflows 25 percent of the time. For this not to be exceeded, 16 spaces should be provided for a group of 25 drivers of the sporadic type. (Don't let this 25 percent alarm you; even when the lot overflows, only one or two drivers are usually affected, and there are other lots for them to go to.)

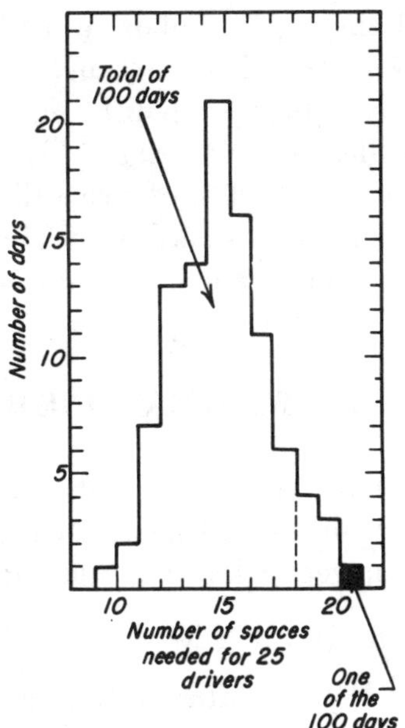

Thus, by simulation, the engineer was able to evaluate the effects of a policy of overassignment (before making any *real* changes) *and* he had something to back up his proposal when presenting it to his client. It took a clerk two days to perform this simulation, employing a simple pencil-and-paper procedure; gathering data and setting up that procedure required another three days. Thus, in five days, it was possible to synthesize 100 days of experience, illustrating the "time compression" capability of simulation.

TECHNICAL TERMS

Write a brief definition for the following term.

Digital simulation

____

____

____

A digital simulation is an experiment with a numerical (i.e., digital) representation of a real-world counterpart, the behavior of which is adequately described by the numerical language.

Digital simulation is experimentation with a digital representation, an "acting-out process" in terms of numbers. It is simple yet remarkably powerful. Because it is a series of step-by-step numerical operations, it can be executed by a computer. This is fortunate. Doing this by pencil and paper is laborious and time-consuming. Actually, digital simulation would be prohibitively expensive in many potential applications if computers were not available to perform these highly repetitive numerical manipulations. Simulation by computer has become popular in engineering, as illustrated by this sample of uses: simulation of the behavior of atomic particles, of space flights, of aircraft movements around an airport, of rush-hour operations of a battery of 95 high-speed elevators in a new skyscraper, of all types of traffic phenomena, and of global warfare.

CASE STUDY: WIRE MACHINE BREAKDOWNS

A machine for twisting strands of wire into a cable must be shut down rather frequently because of breaks and jams. When it does, an attendant must repair the fault before the machine can get back into production. Data is supplied by Tables A and B.

Table A. Data on Intervals Between Wire Breaks

Minutes Between Breaks	Observed Frequency
5	7
10	19
15	28
20	21
25	13
30	9
35	3

Table B. Data on Break Repair Times

Repair Time (minutes)	Observed Frequency
6	12
7	32
8	29
9	17
10	10

APPLY YOUR SKILLS

An engineer is setting up a simulator to determine how many attendants should be assigned to a battery of nine such machines. Set up a Monte Carlo device to generate breaks (actually times between breaks) and a different Monte Carlo mechanism to generate repair times. Explain how the engineer might use these random generators in his model of this battery of machines.

I would determine the breaks (times between breaks) by the following Monte Carlo method.

I would determine the length of repair times by the following Monte Carlo device (make this device different from the previous device).

____

____

____

____

How would you, as the engineer, use these Monte Carlo generators in this digital simulations?

____

____

____

____

"Breaks and jams" are generated by sampling randomly from the data of Table A. Five acceptable methods of doing so are numbered slips of paper, pie chart and spinner, pie chart and random numbers, conversion table and random numbers, or cumulative frequency histogram with random numbers. The cumulative frequency histogram is used thusly:

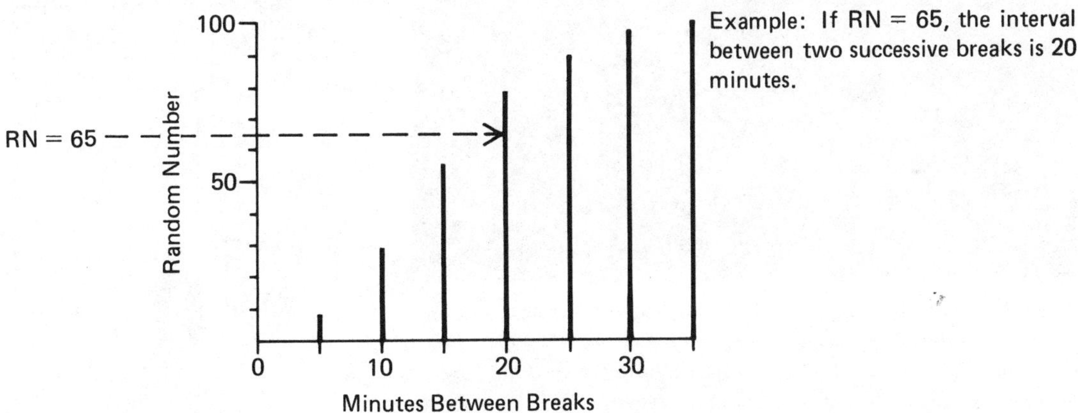

Anyone of those 5 methods can be employed to sample randomly from Table B; however, it must, to comply with the instructions, be a method other than the one selected to generate "breaks and jams."

To model this system, the engineer would assume a certain number of machines to be assigned to an attendant, say two. He could then set up a tabulation sheet on which to keep track of the attendant and waiting machines, using the breakages and repair times he generated by Monte Carlo methods. In this manner, he could generate sufficient operating experience for his first alternative. Then he would do likewise for another number of machines assigned to an attendant, repeating this process until he located the optimum. (You could bring up the matter of the relative costs of attendant and machine and the significance of these.)

A more realistic manner of approaching this problem (but one that is messy for beginners) is to work with all nine machines at once, thus: two attendants assigned to the nine machines, three attendants assigned to the nine machines, etc., with each attendant free to service any machine.

Participative Simulations

It is possible to involve humans directly in a simulation. Picture a group of air traffic controllers in a room. In front of each controller is a radar screen on which the usual pips appear, representing the positions, identities, and altitudes of planes in his area of control. Each controller is in radio communication with these planes, issuing instructions to and receiving reports from planes in his zone of responsibility. To the casual observer, it all looks like the real thing: a busy traffic control center. Yet the "pilots" with whom these controllers are communicating are in the next room (Figure 12).

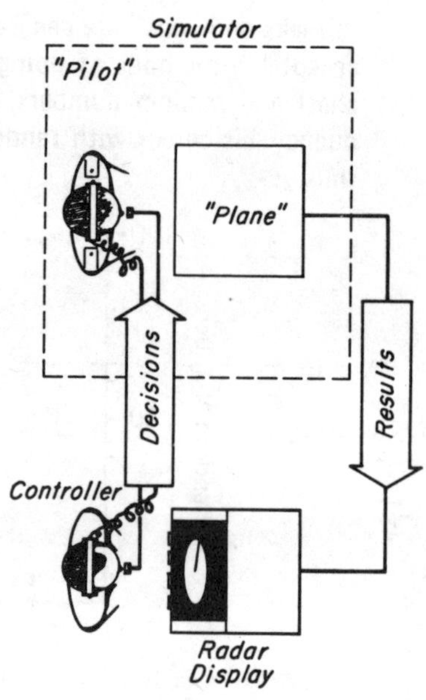

Figure 12. Pilots and planes? Well, not really, but as far as the traffic controllers in the next room are concerned, they are. Each person in this room is playing the part of a pilot, and the special apparatus in front of her is a "plane." As she "flies" her flight plan (e.g., destination, course, altitude) by manipulating the controls on the console, this information is sent by wire to a controller's radar set, where it causes a pip to move realistically across his screen. The controller-pilot conversations take place over telephone lines between the two rooms.

These "pilots" and "planes" constitute a large simulator used by the team of traffic controllers in the next room to test new traffic control procedures. The system operates like this. On the basis of the traffic situation in the controller's zone of jurisdiction, as it appears on his radar screen, he makes decisions and communicates them to the appropriate "pilots." The resulting maneuvers of their planes become apparent on the controller's screen. The controller continues with his decisions and instructions, and the pilots continue to respond, creating a situation that is very realistic, at least for the controllers.

THINK IT THROUGH

Can you provide a brief description of the interaction between a human operator and a simulator?

____

____

The answer to this exercise appears in the following paragraph.

Many simulations directly involve people as decision makers, following the same pattern: the human participant (e.g., a pilot in a trainer) communicates decisions (the pilot through his controls) to the simulator, which determines and communicates the results of those decisions through visual displays, as a continuous cycle. Many driver trainers work on this principle. Figure 13 and 14 provide other examples.

Figure 13. This astronaut, shown maneuvering his spacecraft during rendezvous with another vehicle, can go home at 5 P.M., with other employees of this aerospace company because his is performing this mission on a simulator in the laboratory. The moon and spacecraft images visible to him are being projected onto a large, semicircular screen in front of his "vehicle." As the astronaut manipulates his controls, the scene in front of him changes appropriately, providing him with the illusion of movement with respect to the moon and to the other vehicle. In an adjacent room a TV camera, focused on a three-dimensional model of the moon, picks up the image that appears on this large screen. The camera moves in response to the astronaut's manipulation of the controls. A similar closed-circuit TV system projects the vehicle on the screen. Launches, landings, and orbital missions can be simulated in order to learn what astronauts can and cannot do and to train them for the real thing.

Figure 14. This is a model of a modern supertanker employed to train pilots in the handling of such vessels. The "bridge" has all the essential instruments, and realistic time lags are built into the controls. (It is mainly these long ship-response times that make piloting of large ships such a challenge.) Water facilities at this site enable trainees to practice docking and other maneuvers under normal and emergency situations. A wave-making machine creates the equivalent of 6-meter waves.

Since a human being participates directly, this kind of simulation is termed *participative* simulation. If two or more persons compete with one another in a participative simulation, it is usually referred to as *gaming*. Participative simulation is useful for prediction and for training purposes. (It is preferable to have a fledgling pilot make his learning mistakes on a simulator!)

Thus, an engineer can experiment on iconic, analog, or digital representations in order to make predictions about their real-life counterparts. This process, simulation, is a means of synthesizing experience by operating a representation for a period of time to learn how the real thing will perform.

THINK IT THROUGH

Can you think of any ways that simulation is better than working with real-life counterparts?

____

____

____

____

Answers to this exercise appear in the following paragraph.

Simulation often beats experimenting with the real thing in several respects; it costs less, takes less time, enables engineers to exert closer control over their experiments, and is less hazardous.

MODELS

The representations we have described are usually referred to in the literature and conversation of engineers and scientists as *models*. So we ordinarily speak of graphic models (instead of graphic representations), diagrammatic models, and so forth. This requires that you extend your interpretation of the word model beyond ordinary usage. To the engineer and scientist, a *model* is anything used to describe the *structure* or *behavior* of a real-life counterpart. Models do this through words, numbers, special symbols, diagrams, graphs, or by looking like or behaving like the real-life counterparts that they represent.

TECHNICAL TERMS

To be sure that you understand its meaning, write a brief definition of the following terms.

Model

___

___

___

For a definition, reread the last paragraph.

It may take awhile for you to appreciate fully the generality of this concept. When you do, it will be clear that all of the following are models in the sense that this term is used by scientists and engineers: a mental image of a person or of an experience or of anything; our conception of the nature of light; Darwin's theory of evolution, Einstein's relativity theory, *any* theory; the words dog, stick, book, stone, and most other definite nouns; a verbal description of the workings of a mechanism; a muscial score; and a chemical formula. Then you will not be puzzled if you hear someone refer to Darwin's model, the wave model of light, or your model of engineering.

The term *model* is used universally in engineering and physical science; it is also becoming common in the language of social and biological scientists. So, for communication with persons in these fields, it helps to know what they mean when they talk about the so-and-so model, since they will invariably assume that you are accustomed to using the term model in a broad sense. But understanding this is important for more than terminology reasons. The ability to employ models is a powerful skill that you can almost certainly put to use, regardless of your occupational specialty.

IN SUMMARY

Many models (mathematical equations, three-dimensional models, graphs and diagrams) are certainly familiar enough to you. In these instances, the important difference is the point of view: that these tools have something in common as expressed in the term *model*. A model such as digital simulation, an elegant technique to be sure, is amazingly useful yet surprisingly simple. Of course, digital simulation by computer can't be old hat—computers have not been around that long.

CASE STUDY: THE COMING OF THE SUPERTANKERS

The management of an oil company is contemplating a modification of one of its tanker unloading facilities at a cost of several million dollars (Figure 15). However, some of the consequences of this change are not evident. Therefore they have hired you to answer questions such as these:

- What effect will the elimination of one "standard berth" have on the average time tankers spend waiting for unloading?

- Should standard tankers be assigned to the "superberth" when it is the only place available to unload at the time?

This would be a trivial problem if tankers arrived at predictable times and unload times were uniform. But this is not the case (Figures 16-18). Thus, predicting the effects of the proposed changes is not that simple.

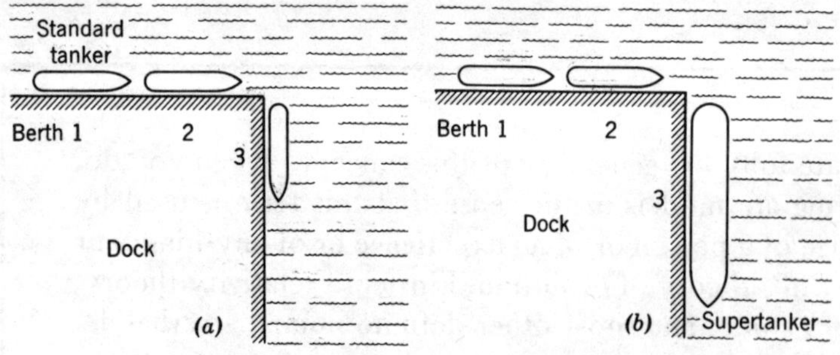

Figure 15. (a) At present the company can accommodate three standard tankers at its unloading facilities. (b) The proposed modification will enable the company to unload supertankers at Berth 3.

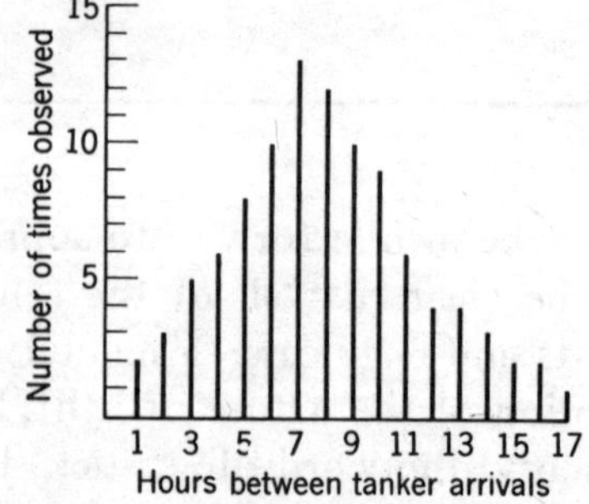

Figure 16. A graphic model of a sample of times between tanker arrivals. These values were obtained directly from 100 successive tanker arrivals as recorded in the dockmaster's log. This, therefore, is past experience in tanker arrivals, which you assume will hold under the proposed setup.

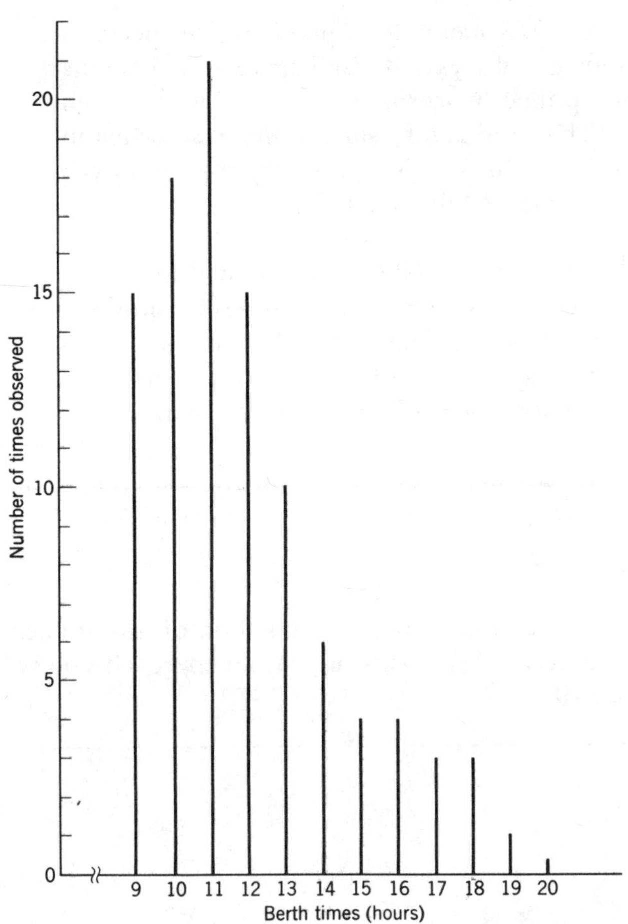

Figure 17. A graphic summary of berth (unload) times for 100 standard tankers, obtained from the dockmaster's log.

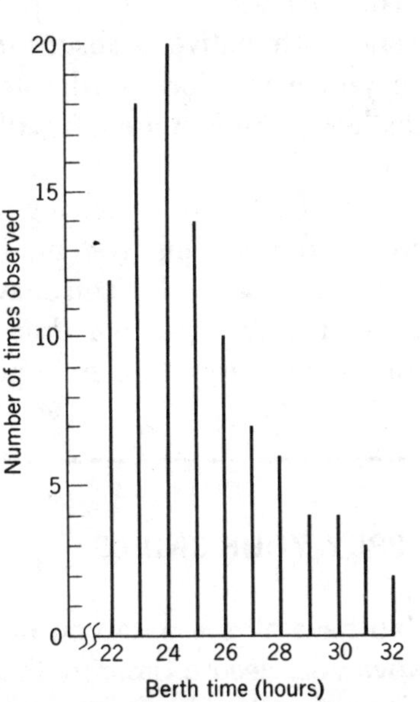

Figure 18. A graphic summary of berth times for 100 supertankers, obtained at another facility this company operates for unloading supertankers. (An important assumption, which seems reasonable since the pumping equipment will be identical, is that this sample satisfactorily characterizes unloading times at the proposal facility.)

THINK IT THROUGH

Solving this problem is not going to be simple. First you must choose a method of predicting the performance of the proposed facility. What methods generally are available and why would you reject one and choose another?

____

____

____

____

____

An answer to this exercise appears in the following paragraph.

Here, as in most such instances, you have at least four methods of predicting the performance of the facility proposed: judgment, building and trying it, mathematics, and simulation. The first will not be accepted *in this case* primarily because the investment is so high. The second alternative is absurd *in this case*. The third is infeasible *in this case* primarily because you would have to develop the equations (which would prove to be prohibitively complex here). So you are left with simulation, digital simulation, in fact.

One of your first steps in setting up a digital simulation model would be to collect data on behavior of the existing system. This has been done for you, with the results shown in Figures 16, 17, and 18. You may safely assume that this data applies to the system proposed. You have been told that 1/6 of the arriving tankers will be "supers." That is all the data you need to set up and operate a digital simulation of the proposed system.

APPLY YOUR SKILLS

You have all the data you need to set up and operate a digital simulation of this problem. Now you need a basic methodological approach. Try diagramming the approach you will take before looking at the answer in Figure 19.

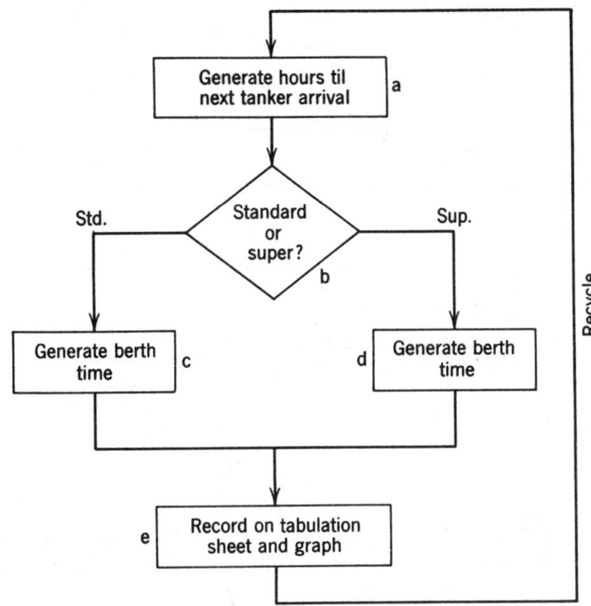

Figure 19. A way of setting up the tanker simulation: Steps a, c, and d require random selection of values from the data supplied by Figures 15-17. Try your hand at each of the Monte Carlo methods introduced in the Appendix at the end of this learning segment. Which one you use where is up to you. The tabulation sheet and the graph referred to at point e are alternative means of keeping a running record of events; try your hand at both. Suggestions for these are made in Figure 20 and 21.

	Tabulation sheet												
(a)	Time between arrivals	2	4	15	4								
	Arrival time	2	6	21	25								
(b)	Tanker type	Su	St	Su	St								
	Start berth 1 (Standard tankers only)		6										
	Start berth 2 (Standard tankers only)				25								
	Start berth 3 (Supertankers only)	2		28									
(c) or (d)	Berth time	26	11	25	12								
	Berth 1 Ready		17										
	Berth 2 Ready				37								
	Berth 3 Ready	28		53									
	Wait time	0	0	7	0								

Identification of experiment: Two standard berths, one superberth. "Standards" NOT allowed at superberths.

Figure 20. Digital simulation tabulation sheet for tanker unloading operations. The lettered rows key them to Figure 19. We have generated the first four tanker arrivals, using Monte Carlo methods to generate times between arrivals and berth times, and a die to make the "standard versus super" choice. The event times for the first four arrivals are plotted in Figure 21.

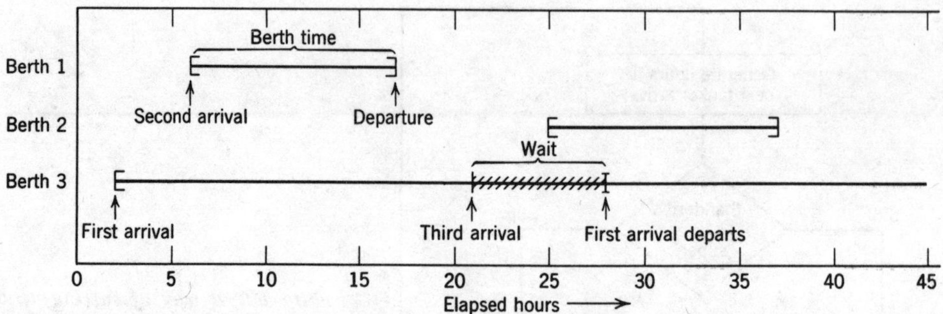

Figure 21. A graphic record of the arrival, berth, and departure times for the first four columns of Figure 20.

YOU DESIGN IT

To generate a significant amount of experience would require that you operate your model for longer than you want to spend. Therefore operate your model only long enough to complete the tabulation sheets provided by the following pages. Do so for each of these experiments.

(a) A facility with two standard berths and one superberth (Figure 15b), under a policy that permits standard tankers to unload at the superberth when the two standard berths are already occupied.

(b) The same facility as in part a, but under a policy that requires standard tankers to unload at standard berths.

After running these two experiments long enough to complete the tabulation sheets, total the hours tankers were forced to wait for unloading. Under the assumption that the company loses $200 for every hour a standard tanker is idle and $700 for every hour a supertanker is idle, calculate the total cost of waiting for policy *a* and for policy *b*. *To repeat: these figures are not statistically significant; 10 days of "experience" is not enough for practical decision-making purposes in this case.*

TABULATION SHEET (Tanker Unloading Operations)

Arrival no.	Arrival time	Tanker type	Start berth 1 (no "supers")	Start berth 2 (no "supers")	Start berth 3 ("supers" and standards)	Berth time	Berth 1 ready	Berth 2 ready	Berth 3 ready	Wait time
1										
2										
3										
4										
5										
6										
7										
8										
9										
10										
11										
12										

Identification of experiment: Two standard berths, one super berth. Standard tankers may unload at super berth.

TABULATION SHEET (Tanker Unloading Operations)

Arrival no.	Arrival time	Tanker type	Start berth 1 (no "supers")	Start berth 2 (no "supers")	Start berth 3 ("supers" and standards)	Berth time	Berth 1 ready	Berth 2 ready	Berth 3 ready	Wait time
13										
14										
15										
16										
17										
18										
19										
20										
21										
22										
23										
24										

Identification of experiment: Two standard berths, one super berth. Standard tankers may unload at super berth.

TABULATION SHEET (Tanker Unloading Operations)

Arrival no.	Arrival time	Tanker type	Start berth 1 (no "supers")	Start berth 2 (no "supers")	Start berth 3 ("supers" and standards)	Berth time	Berth 1 ready	Berth 2 ready	Berth 3 ready	Wait time
1										
2										
3										
4										
5										
6										
7										
8										
9										
10										
11										
12										

Identification of experiment: Two standard berths, one super berth. Standard tankers *not* allowed at super berth.

TABULATION SHEET (Tanker Unloading Operations)

Arrival no.	Arrival time	Tanker type	Start berth 1 (no "supers")	Start berth 2 (no "supers")	Start berth 3 ("supers" and standards)	Berth time	Berth 1 ready	Berth 2 ready	Berth 3 ready	Wait time
13										
14										
15										
16										
17										
18										
19										
20										
21										
22										
23										
24										

Identification of experiment: Two standard berths, one super berth. Standard tankers *not* allowed at super berth.

SELF QUIZ

1. Define the term *model*.

2. Define the term *iconic representation*.

3. Define the term *diagrammatic representation*.

4. Define the term *graphic representation*.

5. List three important uses for mathematical representation.

6. Define the term *iconic simulation*.

7. Define the term *analog simulation*.

8. Define the term *digital simulation*.

9. Define the term *participative simulation*.

10. Define the term *gaming*.

____

____

11. List four reasons why simulation may be preferred to operating the real thing.

____

____

____

____

1. A model is anything used to describe the structure or behavior of a real-life counterpart. Models do this by means of words, numbers, special symbols, diagrams, graphs, or by looking like or behaving like the real-life counterpart.

2. An iconic representation is a model that looks like some real-life counterpart, either in two or three dimensions.

3. A diagrammatic representation is a model, consisting of a set of lines, symbols, and numbers, that conveys the structure or behavior of some real-life counterpart.

4. A graphic representation is a model, consisting of lines, numbers, and other symbols, that conveys the relationships between relative magnitudes of the components of some real-life counterpart.

5. Mathematical representation is very helpful for prediction, communication, and reasoning.

6. An iconic simulation is an experiment with a representation of a real-life counterpart that looks like the counterpart, i.e., an experiment with an iconic representation.

7. An analog simulation is an experiment with a medium that behaves in a manner similar to a real-life counterpart.

8. A digital simulation is an experiment with a numerical (i.e., digital) representation of a real-life counterpart.

9. A participative simulation is an experiment with a model that allows a subject to interact with the model and in which the model reacts to the decisions made by the subject operating it.

10. Gaming is an experiment involving several subjects interacting with a model.

11. Simulation may be preferred to working with the real-life counterpart because it costs less, it takes less time, it allows the engineer to have more control over the experiment, or it involves less risk.

APPENDIX: MONTE CARLO METHODS

A barber would like to know, *before* he purchases the equipment and hires the new man, what effects adding a second barber to his shop will have on customer waiting and barber idleness. Digital simulation is an effective means of making the predictions he desires. To set up a digital simulation for him the author proceeded as follows.

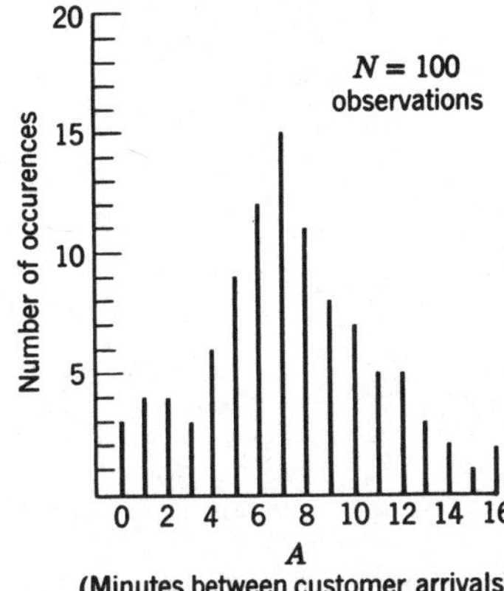

Figure 22. A frequency histogram summarizing the 100 observations made of minutes between successive customer arrivals at the barber shop.

1. I observed the present one-barber operation to collect a sufficient sample of times *between* successive customer arrivals and of haircut times, observing long enough to record such times for 100 customers. The resulting A values (times between arrivals) and C values (chair times) are summarized by Figures 22 and 23.

2. Then I took 100 slips of paper and wrote the 100 A observations on them, which meant there were three slips with zero on them, four slips with 1 on them, four slips with 2 on them, and so forth.

3. I did likewise for C times.

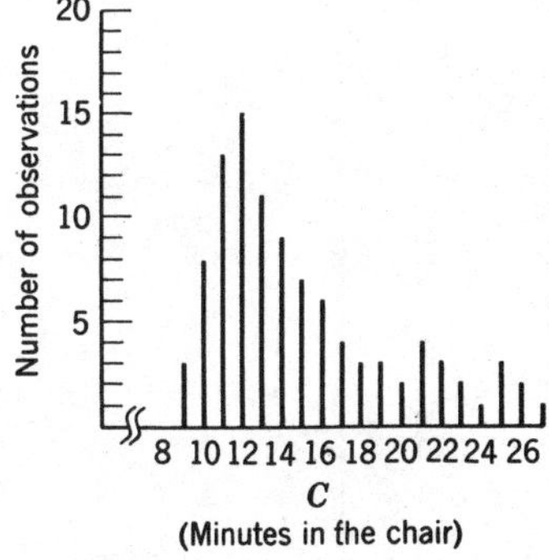

Figure 23. A frequency histogram summarizing the sample of "chair times."

4. I set up a table (Figure 24) to provide a chronological record of events about to occur, as this shop is operated on paper by a process called digital simulation.

5. The simulation model was set in motion. To do so, I had before me the two piles of slips and the tabulation sheet. I assumed for this test run of the model that there was one barber and that the shop opens at 8 A.M. Literally, I picked an A value at random from that pile, and used it to determine the time of the first customer's arrival. The value selected (and returned to the pile upon reading it) was $A = 10$ minutes, and so the time of his arrival was

$$T_a = 8{:}00 + 10 = 8{:}10$$

That fact was so recorded on the tabulation sheet. Of course, since he was the first arrival, there was no waiting, and so $T_a = T_c$, where T_c is the time at which the first customer gets in the chair.

TABULATION SHEET

	Customer number →	1	2	3	4	5	6
A	Value selected	10	15	4	7	13	15
T_a	Arrival time	8:10 (8:00+10)	8:25 (8:10+15)	8:29 (8:25+4)	8:36	8:49	9:04
T_c	Cutting starts	8:10	8:25	8:37 (previous T_e)	8:55		
C	Value selected	14	12	18			
T_e	Cutting ends	8:24 (8:10+14)	8:37	8:55			
W	Customer wait	0	0	8 (8:37-8:29)	19		
I	Barber idle	10 (8:10-8:00)	1	0	0		

Figure 24. Tabulation Sheet for barber shop digital simulation.

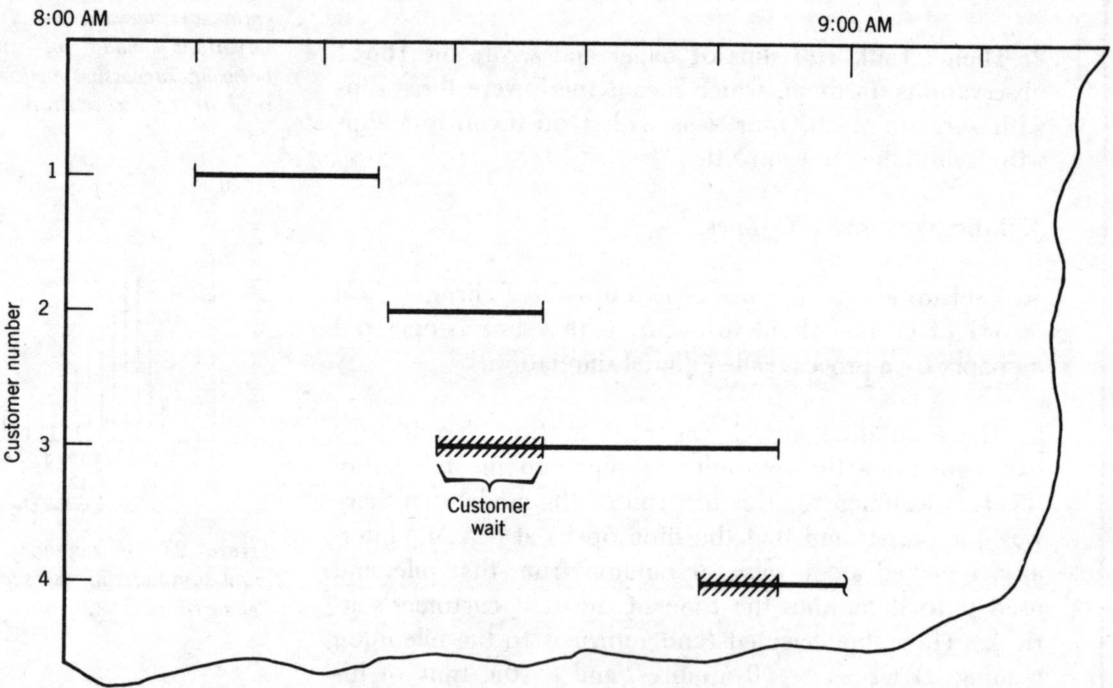

Figure 25. Graphic record of barber shop simulation.

6. In similar fashion, a C value of 14 minutes was selected and recorded in the table and, from that, the time at which the barber was finished with customer one was determined as

$$T_e = T_c + C = 8{:}10 + 14 = 8{:}24$$

These times can be found on the graph of Figure 25.

7. That procedure was continued until enough experience had been synthesized to justify conclusions. Notice from Figures 24 and 25 that customer waiting occurred occasionally. At the end of the experimental run, these wait times were summed, and so were barber idle times.

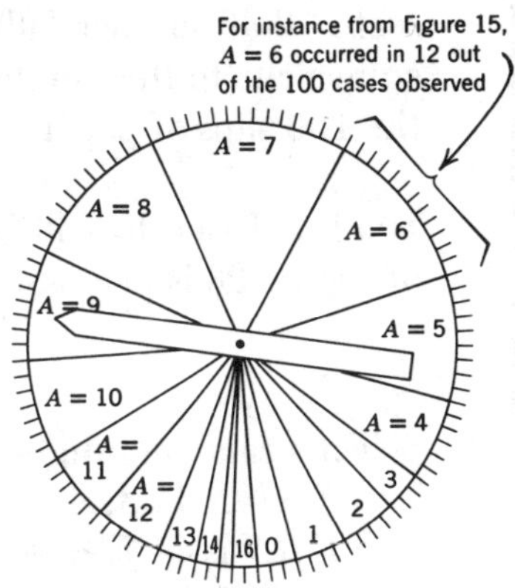

Figure 26. Pie chart equipped with free-spinning pointer in order to select randomly "time until next arrival."

8. But hold on. There are methods of making these random selections of A and C—called *Monte Carlo methods*—that are more sophisticated and convenient than the slips-of-paper routine. One of them involves a pie chart and free-spinning pointer. One was constructed for generating A values by replotting Figure 22 in circular form. The resulting pie chart (Figure 26) is equipped with a free-spinning pointer and becomes a perfectly acceptable method of choosing values randomly from a population of numbers.

9. But there is a better way, based on a table of random numbers. Picture yourself with a spinner device like that shown in Figure 27, which has equally spaced marks numbered 1-100 around the circumference. You spin the pointer and record its stopping point in terms of this 1-100 scale. The results of many repetitions of this procedure are shown in Table A.

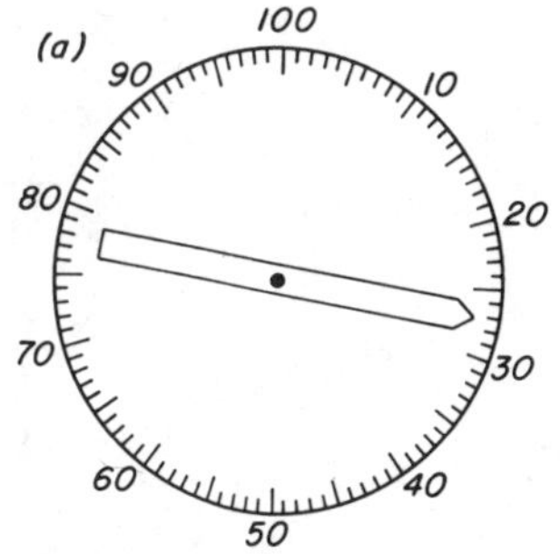

Figure 27.

Table A. Table of Random Numbers

12	82	89	54	11
74	41	21	02	13
81	06	19	79	91
19	43	17	75	82
29	21	35	18	57
47	98	81	96	28
26	60	24	77	49

In *the long run*, each of the numbers 1-100 will appear with equal frequency in a table of random numbers, but in random order. Such tables are readily available in textbooks and handbooks.

This table can replace the pointer by Figure 26. By numbering from 1 through 100 around the circumference of the pie chart, creating Figure 28, it is now simply a matter of selecting a number (you can go across rows, down columns, or pick at random, it matters not) from Table A and consulting Figure 28 to determine in what

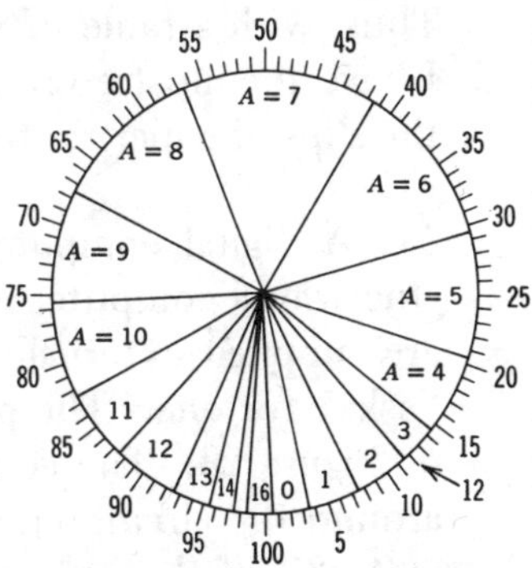

Figure 28. The pie chart of Figure 28 with the spinner replaced by a circumferential scale of 1-100.

sector that number falls. If I select 12 from Table A, that falls in the $A = 3$ sector. This is equivalent to the pointer of Figure 26 stopping in the $A = 3$ sector *and* to selecting a 3 from the 100 slips of paper.

10. But I can do still better. For the procedure described in step 9, not even the pie chart of Figure 28 is necessary. The whole process can be reduced to a series of rules, as follows.

Select a random number (*RN*) from Table A. Then:

IF RN is 1–3, $A = 0$ minutes
IF RN is 4–7, $A = 1$ minute
IF RN is 8–11, $A = 2$ minutes
IF RN is 12–14, $A = 3$ minutes
.
.
.
IF RN is 99 or 100, $A = 16$ minutes

These rules can be conveniently summarized into what is commonly called a *conversion table* (Table B).

Table B. Conversion Table

IF random number (*RN*) is:	*THEN* time until next arrival (*A*) is:
1-3	0 minute
4-7	1 minute
8-11	2 minutes

Thus, with a table of random numbers (1 through 100 in this case) and a conversion table, I have the most convenient of the "pencil-and-paper" Monte Carlo methods. This method, the slips of paper routine, and the pie chart and spinner mechanism are equivalent.

11. A digital computer can also make random selections from a population of numbers. One way a computer can do this is diagrammed in Figure 29. In this case, random numbers are supplied externally; they are on punched cards and available whenever the computer "asks" for one. The procedure diagrammed here can be explained in terms of the pie chart of Figure 28. The computer asks for a random number and then, in effect, works its way around the circumference, starting with the $A = 0$ sector. It asks: "*RN*, do you fall in this sector?" If the answer is yes, it assigns $A = 0$, but if the answer is no, it goes to the $A = 1$ sector. If the answer there is yes, *A* becomes 1 minute; otherwise the machine goes to the next sector. It continues around the circumference, stepping from sector to sector until it scores a "hit" and assigns the appropriate *A* value.

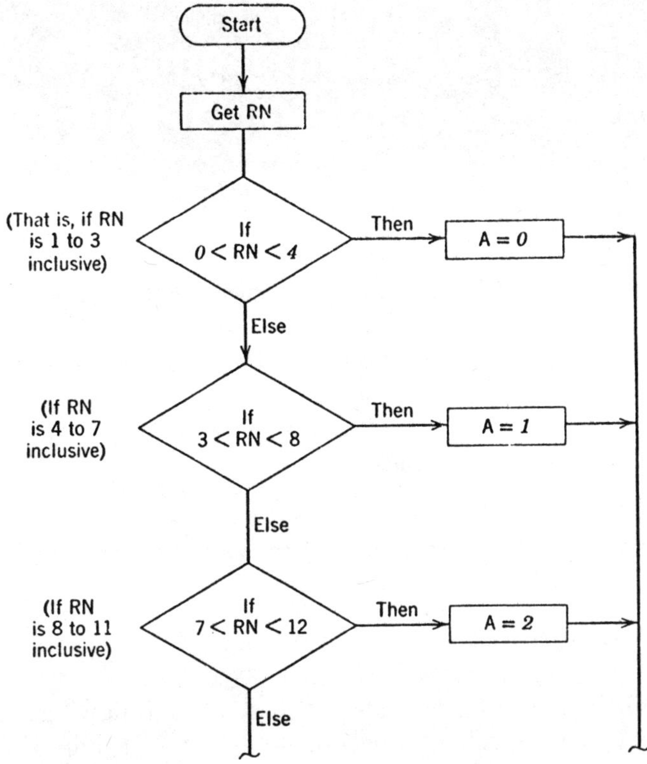

Figure 29. Diagram of a portion of a computer routine for Monte Carlo-ing. If so desired, the computer can generate random numbers internally through a subroutine.

If you understand the procedure just described, and if you are able to program a computer, you are equipped to write computer simulation programs. This is no small matter, for simulation by digital computer is a very powerful technique of rapidly increasing importance.

Learning Segment 8
USING AND DEVELOPING MODELS

LEARNING OBJECTIVES

- Recognize and understand the various uses for models.
- Recognize and understand the essential differences between models and their real-life counterparts.
- Develop and apply predictive models.

HOW ENGINEERS USE MODELS

Models have great utility in engineering, so much so that the following overview of engineers' uses of models tells you much about the profession. In fact, one of the engineer's more conspicuous attributes is his quickness to turn to a model.

For Thinking

A model can be a real help when trying to visualize the nature or behavior of a system or phenomenon that the unaided mind finds difficult to grasp. There are electrical circuits, manufacturing systems, chemical processes, and mechanisms, the complexity of which makes a diagrammatic or other type of model essential to human comprehension. Iconic, diagrammatic, and graphic models are especially helpful in providing a compact, overall, simplified view of the whole. Often, in thinking of a physical phenomenon, the engineer finds it expedient and profitable to think of it in terms of a model. For instance, an experienced engineer usually thinks of an alternating electrical current as a sine wave in graphic form instead of as the movement of electrons in a conductor.

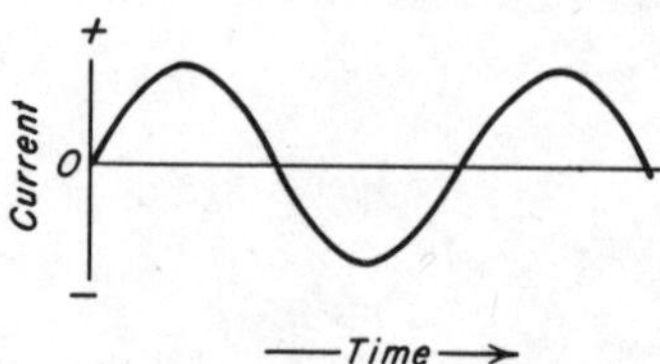

An engineer who designs aircraft or long-span bridges will most likely think of wind gusts in graphic form rather than as chilling blasts in the face. Often it is these abstractions that engineers manipulate in their thinking.

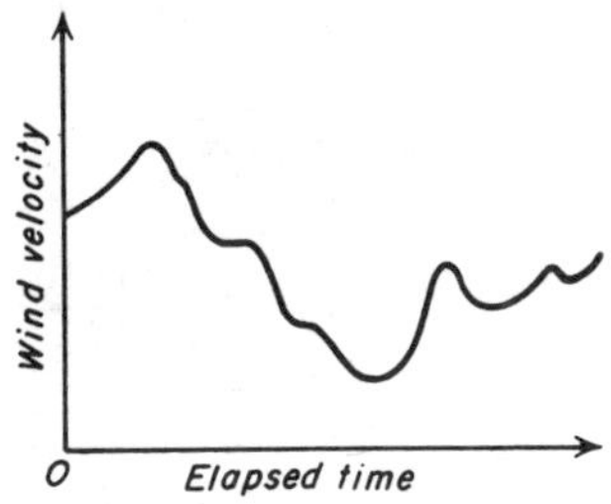

A major objective of an engineer's education is to develop the ability to think of physical phenomena in terms of useful abstractions.

For Communication

Of course, most communication is through models. An author communicates to you through symbols, photographs, sketches, and diagrams. In addition to these common communication aids, engineers frequently use mathematics, graphs, three-dimensional models, and drawings. They rely on models especially when the systems and phenomena to be communicated are complex.

For Prediction

Recall that a major part of decision making is predicting the performance of alternative solutions with respect to the criteria selected. Models are utilized extensively for this purpose; in fact, engineers would be lost without these fundamentally important tools.

Here is an example of the usefulness of predictive models in design. Several of the television transmission antennas atop the Empire State Building are to be replaced (Figure 30). The engineer in charge thinks it desirable to predict how much the modified antenna stack will bend as a result of wind load. There is no mathematical model specifically for predicting bending (deflection) of such antennas, but he (an electrical engineer) remembers something about the analysis of beams from basic engineering courses. He recalls that mathematical equations for predicting just about anything you would want to know about a beam rigidly fixed at one end (called a *cantilever* beam) are readily available in textbooks. He can treat the antenna stack as a vertical cantilever beam and use this model, which predicts end deflection (d) of a beam uniformly loaded over its length (L).

Figure 30. Eight television stations have their transmission antennas stacked atop the Empire State Building. Several of these, near the top, are to be replaced with improved designs. That is a challenging problem.

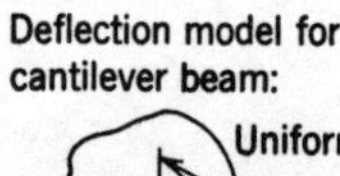

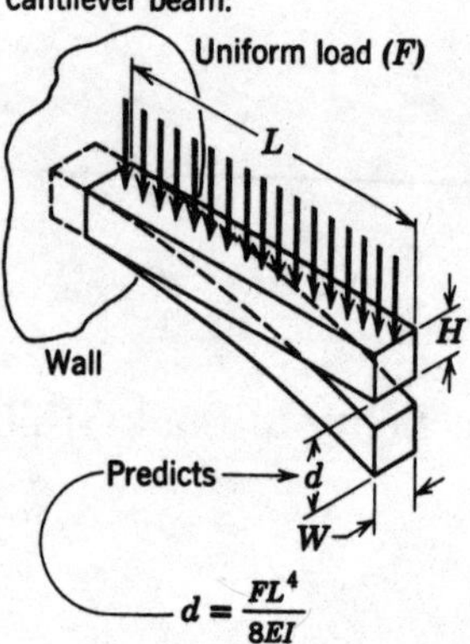

$$d = \frac{FL^4}{8EI}$$

$$d = \frac{280 \times 50^4}{8(2.07 \times 10^{11})(7.866 \times 10^{-3})}$$

$$= .134 \text{ centimeters}$$

THINK IT THROUGH

Now that this engineer has mathematical tools in the form of equations that will help predict the behavior of the antenna, what remains to complete this experiment via modelling?

____

____

____

Some answers to this exercise appear in the following paragraph.

All that remains to predict d is to obtain certain data (e.g., normal and maximum wind velocities at this altitude), to make certain assumptions (e.g., wind resistance of the stack), and then to substitute these values into the deflection equation. You are being spared the details; they are unimportant at the moment. What *is* important is that you appreciate the utility of this model in *predicting* behavior of the structure while it exists only on paper; herein lie the beauty and the power of a predictive model.

Recall the Severn Bridge pictured in Learning Segment 1. During the design of that bridge, extensive wind tunnel tests were made on alternative cross-sectional shapes (20 of them, in fact) of the bridge deck to arrive at a shape that minimizes wind effects.

Something that could be overlooked is the effect of winds during the delicate operation of lifting and positioning the deck sections. Engineers simulated this operation in a wind tunnel and learned that special provisions had to be made to control wind effects during these operations.

The above cases are typical in this respect: in solving problems an engineer must evaluate most solutions while they are still in the conceptual stage. Models are extremely useful for this purpose; they enable him to make the required predictions of solution performance without the necessity of physically creating the solution. Through the manipulation of mathematical and simulation models, it is possible to evaluate solutions with less time, cost, and risk than experimentation with the real thing ordinarily requires, yet more accurately than is usually attainable when pure judgment is used. The engineer designing the antenna stack is not about to have a full-scale version of his design mounted atop the building just to measure deflection! Yet the stakes are too high to rely soley on opinion. Predictive models are an excellent compromise in such situations.

YOU DECIDE

Obviously, no predictive model perfectly represents its real-life counterpart; assumptions are never perfectly satisfied. Describe a realistic attitude toward discrepancies between assumed and true states of affairs.

____

____

____

____

Some discrepancies between the assumed state of affairs (i.e., the model) and the actual state of affairs are uneconomical (if not impossible) to avoid. To put it another way, an errorless predictive model is uneconomical to achieve, even to approach. So some discrepancies should be accepted as natural and inevitable.

For Control

When developing a model for prediction purposes, an attempt is made to have the model's predictions agree as closely as feasible with what eventually occurs. In some situations, the converse is true; an attempt is made to get the modeled situation to conform to the model. The engineering drawings for a building constitute a model and, of course, the building is constructed to conform to that model. The flight path a spacecraft must follow to reach its objective is carefully computed beforehand. This planned flight path is a model; very elaborate systems are employed to hold the spacecraft's actual flight path to that model.

For Training

Most models that are useful for communication are useful for instruction. Not so obvious, however, is the usefulness and growing popularity of participative simulation as a training tool, especially when the costs of blunders are high because of danger to life or expensive equipment.

THINK IT THROUGH

Can you think of some situations for which training via models would be particularly appropriate?

Some answers to this exercise appear in the following paragraph.

Simulation is relied on extensively to train commercial and military pilots, air traffic controllers, and astronauts. The astronauts and all key ground personnel repeat their feat many times through simulation in the laboratory before the real countdown and blastoff. Similarly, the crews that operate our missile defense and attack systems get no opportunity to practice at the real thing, yet they must become proficient at their tasks. The answer is simulation.

An engineer's ability to employ a variety of models for the above purposes is important, indeed, since a wide variety of models serve a multitude of purposes in design—the usefulness of diagrams in structuring problems, the value of sketches in creative thought, the role of mathematics and simulation in predicting the performance of alternative solutions, and the usefulness of most types of models for specification of the final solution.

USES OF MODELS IN OTHER FIELDS

It is worth observing, as you probably did for the design process, the applicability of modeling skills to nonengineering problems. An important purpose of citing nontechnical extrapolations of skills such as design and modeling is to enable you to visualize specifically what people are talking about when they speak of the broad applicability of engineering skills.

You know that models are used extensively in science. A book in any branch of physical science, physics, for example, is packed with various kinds of models. This is also true for the biological sciences in which iconic models are conspicuous, and for the social sciences, where verbal models predominate. So this is no news to you. But one aspect of modeling that you do not encounter routinely in textbooks is the variety of applications of digital simulation.

THINK IT THROUGH

We are about to mention a number of ways in which digital simulations can be applied. You should be able to anticipate a few. Try.

____

____

____

Answers to this exercise appear in the following paragraph.

It is possible to simulate digitally: the behavior of neurons and neuron networks; cancer cell multiplication; movements and collisions of atomic particles; learning processes; consumer behavior; business activities of numerous kinds; economic activity at the regional and national level; meteorological phenomena; and global warfare. In fact, these are not only possible, they have been accomplished, in each instance with the help of a digital computer. So digital simulation does have widespread applicability outside of engineering.

Digital simulation by computer may subsequently serve a more important role in the biological and social sciences than it has in physical science and engineering. Chance is a significant factor in the behavior of living things, and the Monte Carlo technique is a means of synthesizing the role of chance. So individual and group behavior of all kinds of living organisms can be realistically simulated by a digital computer. This is especially significant because biologists and social scientists find it difficult to conduct satisfactorily controlled experiments when dealing with the actual phenomenon.

MODEL VERSUS REAL WORLD

Some realities associated with the operation of a campus parking lot were ignored—"assumed away," as the saying goes—by the simulation model described earlier. As far as the model is concerned, weather has no effect on the probability of a driver using his car, yet we know this is not true in real life. Furthermore, there are some features of the parking lot model that are not true of the real thing. For example, it was possible in the model to generate a negative park time (it was ignored if it occurred), which obviously doesn't occur in real life. Such discrepancies between the model and its real-world counterpart are inevitable. They are found in *every* model.

There are such discrepancies in the case of the model used to predict deflection of the Empire State antenna stack; for instance:

- According to the model, the beam is constructed out of homogeneous material, yet it is welded out of steel angles.

• The wind load is assumed to be constant, or at least gradually applied, yet the winds especially at that level are very gusty.

• The model is based on a beam of uniform cross section over its length, yet the antenna stack tapers down in stages.

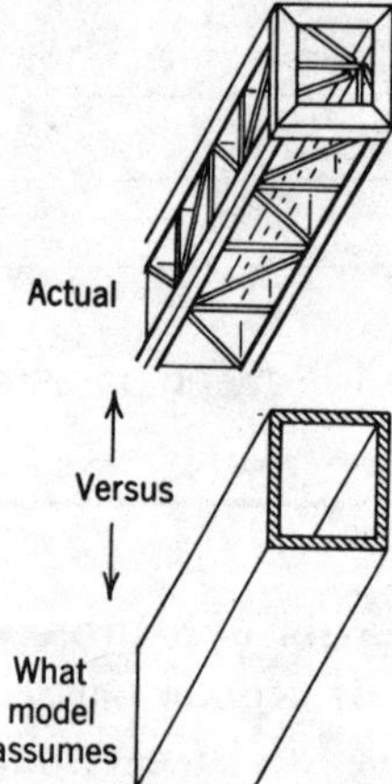

Yet these models can provide useful predictions. The fact that some assumptions made by a model are not consistent with reality, by itself, is unimportant. What matters in predictive modeling is whether or not the final result—the predictions—are satisfactory for the particular purpose at hand.

THINK IT THROUGH

Why do you suppose it is necessary when developing and using models to make simplifying assumptions?

____

____

One answer to this exercise appears in the following paragraph.

Simplifying assumptions are made for good reason. If some of the complicating factors are not excluded from the model, it is virtually impossible to develop a usable predictor. Furthermore, many discrepancies between model and reality are of negligible *practical* consequence; their removal makes a model more complicated and costly and yet does little to improve accuracy. To be sure, some of the discrepancies between the parking lot model and the real thing could be removed, but the model would take much longer to set up and run; the small increase in accuracy doesn't justify it. In every instance, complicating, costly factors that are irrelevant or inconsequential are ignored, and understandably so.

DEVELOPING PREDICTIVE MODELS

The following procedure for developing a satisfactory predictive model is a significant one in science and engineering.

1. **A model that is *potentially* satisfactory for the prediction task at hand is developed or selected.** It can be a model prepared especially (e.g., the model for simulation of parking lot operations), or it can be one selected from the store of "ready-made" models, as in the case of the deflection equation. In the latter case, the engineer on that job is apprehensive about the violated assumptions, to the point where he questions whether that model is accurate enough in this application. Therefore, before he uses the deflection model in this and future antenna design problems, he intends to evaluate its predictive ability.

2. **Observations are obtained from the "real-world" with which this model's predictions can be compared.** In this instance, the observations of what actually occurs are obtained from realistic laboratory tests. A 10-foot scale model of the antenna stack is subjected to simulated wind loads of varying intensity, gustiness included, so that deflection can be measured. For the same experimental conditions, deflection is *calculated* from the deflection equation. Thus, for each set of conditions tested, he has a *predicted* deflection and *observed* deflection, which he plotted as shown in Figure 31a. The poorer a model's predictive ability is, the weaker the correlation between predicted and actual results is, and the more scattered the points in a plot like this are.

3. **Observations are interpreted and a decision is made whether to use the model as is, refine it and repeat the process, start over with a different model, or give up on models and turn to another means of making the predictions needed.** In general, this decision depends heavily on the situation in which the predictions are to be used. When the costs of errors are high, especially when life and limb are at stake, the predictions must be very accurate. In other situations, fairly rough predictions are adequate. Hence, the adequacy of a model cannot be evaluated independently of the particular use to which it is to be put. Therefore, one should not attempt to judge the seriousness of the dispersed points of Figure 31a unless one knows specifically how and where the model's predictions are to be used.

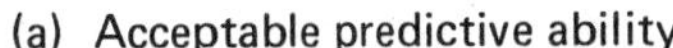

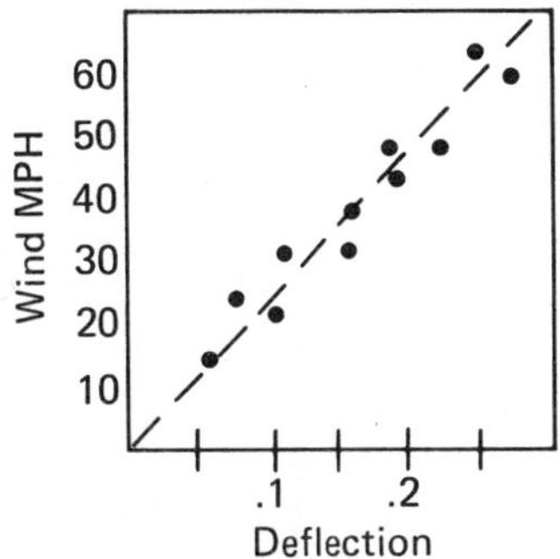

(b) Perfect predictive ability

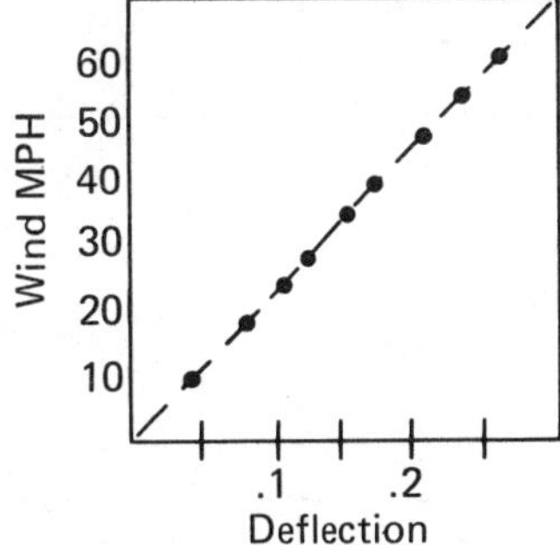

(c) No predictive ability

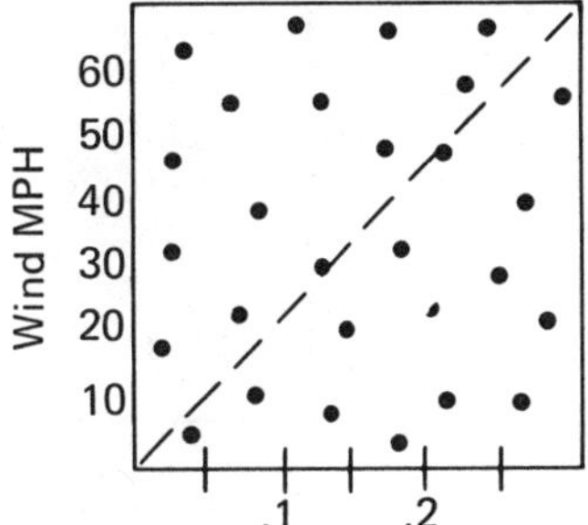

Figure 31. Dots represent actual measurements of deflection; broken line represents model's predicted deflection.

LET'S BE SURE ABOUT THIS

You have all the information you need in the preceding three paragraphs to fill in the blank spaces in the following simplified diagram of the procedure for developing a predictive model.

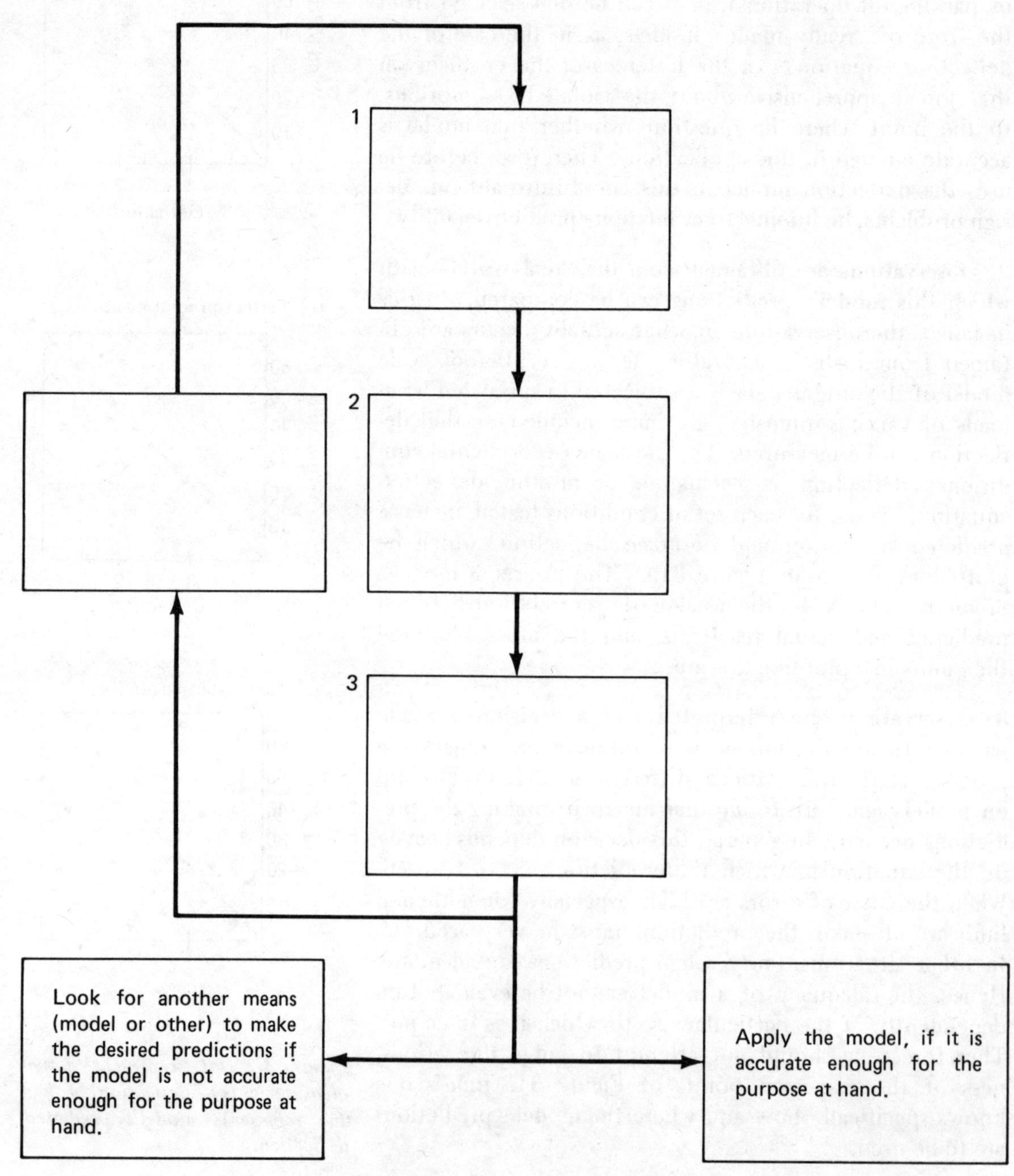

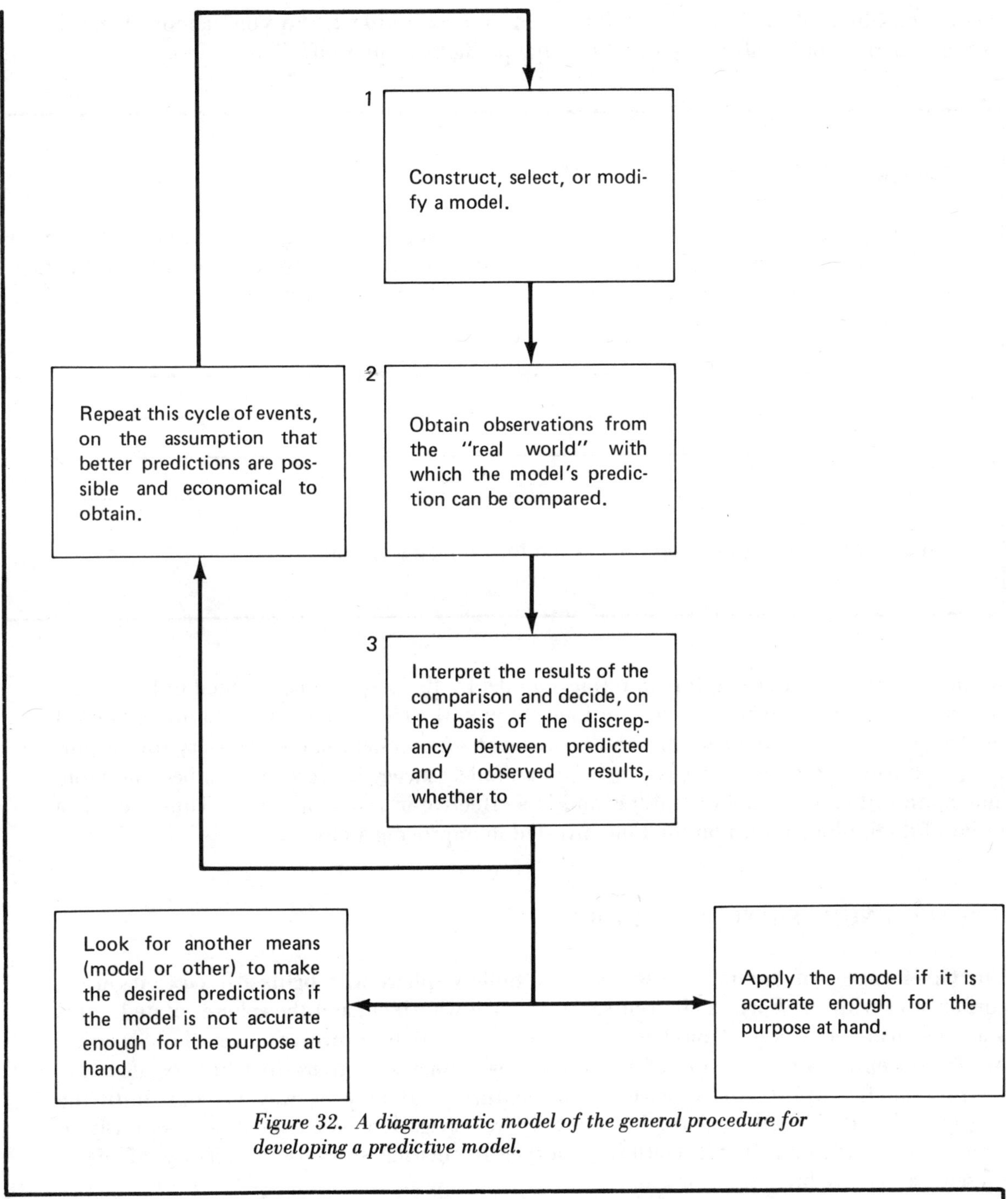

Figure 32. A diagrammatic model of the general procedure for developing a predictive model.

You cannot expect the perfect predictive ability demonstrated in Figure 31b for several reasons. The points will never fall on a straight line because assumptions will inevitably be violated to *some* degree whenever a model is applied. Errorless predictions are unattainable. Furthermore, there is some error in the measurement of the values substituted into the model, for example, F and E. Note, too, that predicted deflection is compared with *mea-*

sured (i.e., observed) and not actual deflection of *a scale model*. So you cannot blame all of the dispersion of points in Figure 31a on the predictive equation.

THINK IT THROUGH

This procedure for developing a predictive model allows for looping back to repeat the three basic steps in order to refine the model. But there is a limit to the amount of refinement one is willing to attempt for such a model. What do you suppose is the primary reason for this limit?

____

____

____

____

An answer to this exercise appears in the following paragraph.

As more time is spent attempting to refine a model by the process diagrammed in Figure 32, the development cost of the model continues to mount. The engineer is certainly interested in this cost. He will invest no more in refinement of a model than is necessary for his purposes. Furthermore, as he works to refine a model, successive refinements become more and more difficult to achieve and less and less effective in reducing error. Thus there is a point of diminishing return on the time invested in improving a model.

SCIENCE, ENGINEERING, AND MODELS

The Greek model of the universe was a large, hollow sphere with lights affixed to its inside surface and with the Earth at its center. The Earth was fixed, and the sphere rotated. This was one in a succession of models of the universe that have evolved over the centuries. Man's conception of the shape of the Earth has a long and interesting history also. Of course, the flat-world model persisted for centuries. It may be way off but, until the voyages of sailors lengthened to the point where navigation based on it was seriously in error, the discrepancy didn't matter. Another fascinating story is the history of man's models of his own brain.

The history of science is replete with cases like these of models that (we know now) were very unrealistic representations of reality. Even today discrepancies exist in all scientific models, of lesser magnitude and consequence to be sure. The history of science is essentially the history of man's attempts to remove discrepancies from his models of nature. In fact, science *is* the continuing process of refining models of the natural world.

THINK IT THROUGH

Both the sciences and engineering utilize models extensively. But there are differences in what is emphasized in the sciences and what is emphasized in engineering in relation to models. Can you think of any differences between the way models are treated in the sciences and the way they are treated in engineering?

____

____

____

____

Answers to this exercise appear in the following paragraph.

If you are clear on the differing roles of models in science and engineering, you are clearer on the distinction between the two fields than some writers are. To the engineer, a model is a means to an end, a tool. But in science, the model *is* the end. Furthermore, on the matter of when and how long to strive for refinement of a model, there is a significant difference between science and engineering. The engineer's basis for deciding when to terminate the refinement cycle diagrammed in Figure 32 is basically an economic one; as an engineer, one invests no more in a model than necessary to solve the problem. A scientist has other bases for deciding when to terminate this process that are less tangible, quite elusive, and often personal. In many instances he is driven by personal interest in a subject, by dedication to a cause, by an obsession, or by an urge for perfection, to go far beyond the point an engineer would go in refining a model. This is not to say that a scientist is oblivious of matters such as the cost of continued efforts to refine a model, of diminishing returns on those efforts, and of other scientific questions he could turn to, but the weights attached to these criteria by scientists and engineers are certainly different.

Modeling—Finale

This concludes two learning segments on modeling but hardly terminates your contact with models. In fact, in the next learning segment you will gain additional insight into the roles of models. If you are familiar with the types of models, and if you have developed a feel for the utility of models, you have assimilated something that you will almost certainly find useful on numerous occasions.

CASE STUDY: A MODEL FOR CATERERS

A large hotel caters to all kinds of banquet affairs, ranging from small receptions to large conventions. The banquet manager has a problem. If he "overprepares" food, then set-up time is wasted; if he underprepares, he has problems, too. So he is interested in accurate forecasts of turnouts. Actual turnout depends on the type of affairs and numerous other variables.

YOU DESIGN IT

1. Outline the steps you would take in order to develop and validate a predictive model for the manager. Include a list of variables that you feel *may* have to be recognized by this model.

2. Perhaps you can contact a local caterer and work with this firm to actually carry out the steps for developing and validating a predictive model in this case.

This is a practical problem. You may well find a local caterer who will cooperate and benefit from working out a solution to it. These are the steps you might take:
(a) Identify potential predictors (i.e., measurable variables on which turnout depends). Some of these are

- Number of persons estimated (based on ticket sale, reservations, customer's estimate, etc.)

- Weather

- Type of affair (wedding reception, retirement party, etc.)

- Payment method (advance ticket sale, at-the-door, or lump-sum payment)

(b) Decide what variables to investigate.
(c) Collect data (hopefully the caterer's records will suffice).
(d) Analyze data and formulate tentative model.
(e) The remaining steps are as diagrammed in Figure 32.

CASE STUDY: GETTING THE PRODUCT TO MARKET

We have stressed that a number of engineering techniques are relevant to problems that you do not ordinarily associate with engineering. Here is an opportunity to demonstrate this to yourself. Set up and run a digital model with which you can simulate operation of the system outlined in the following diagram (Figure 33).

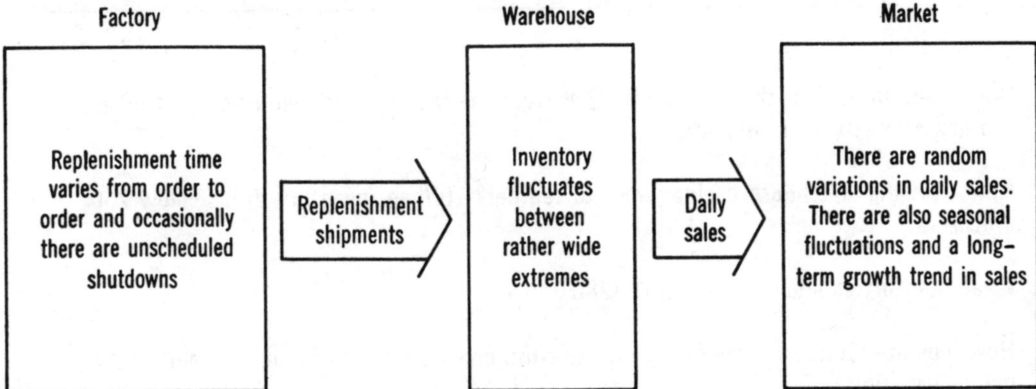

Figure 33. Representation of a warehousing system showing some of the real-life behavioral characteristics of each component. The resulting inventory fluctuations characteristic of all inventory systems are characterized by Figure 34.

What are some of the questions for which you will need answers before you can construct and utilize a model for this case? You will find it useful to first consult Exhibits A–C.

How many units does the warehouse order from the factory when replenishment of inventory is necessary? (150 units)

How does the warehouse decide when to reorder? (When inventory drops below 250 units.)

What inventory should we start with? (280 units)

How long does it take for the factory to replenish inventory? (See Exhibit A. Notice the natural variation.)

How many units are sold per day? (Exhibit B)

Are there seasonal trends in sales, or other cycles? (Yes, but you may ignore them in your model. However, you might think about how a seasonal cycle could be incorporated into the model.)

Exhibit A:

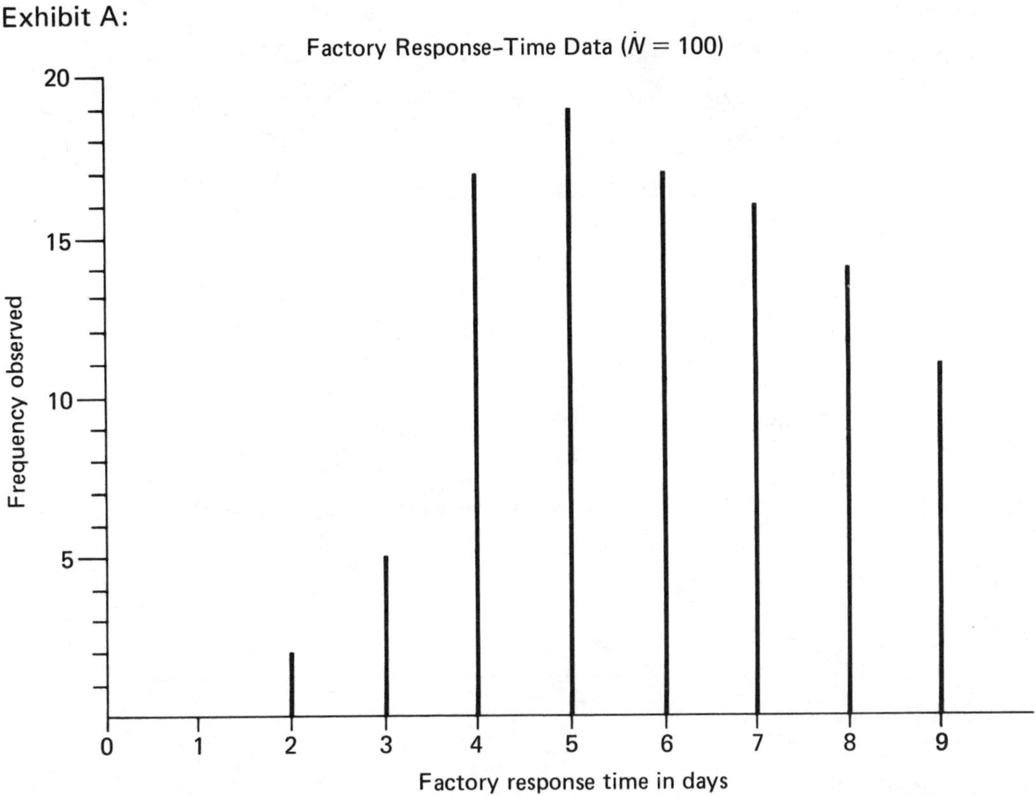

Exhibit B:

Sales Data

Units Sold Per Day for a Sample of 100 Working Days

11	12	16	14	15
7	15	17	9	12
4	13	12	7	13
12	6	9	16	10
8	7	14	19	8
16	15	15	13	5
13	5	11	15	4
8	12	10	17	12
11	11	13	13	11
9	17	15	14	8
9	18	19	11	9
7	15	9	11	11
13	6	8	9	14
10	10	12	11	17
13	12	14	15	16
8	14	16	16	10
9	18	10	8	10
10	11	12	6	11
14	10	9	5	17
14	7	13	16	12

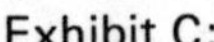
Exhibit C:

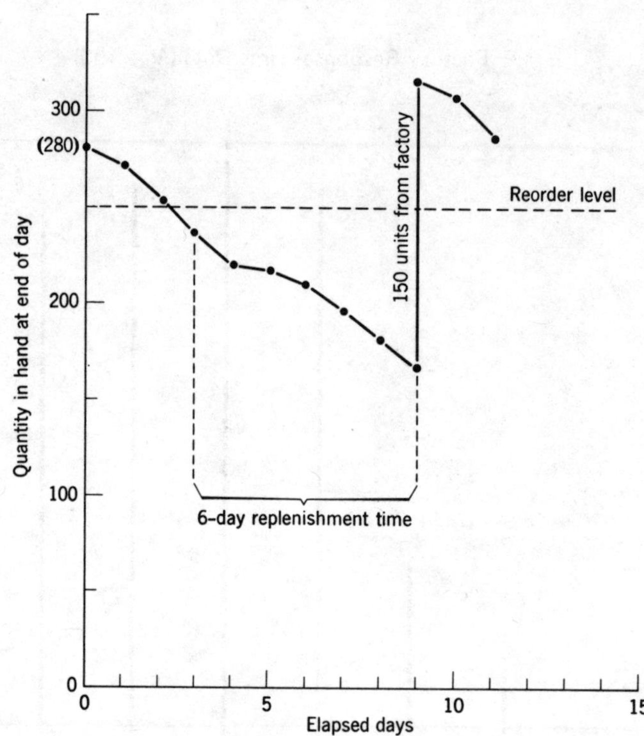

YOU DESIGN IT

As a minimum, develop Monte Carlo mechanisms (employing a different method in each instance) that generate daily sales and introduce real-life variation in factory replenishment time (which is the time lapse between the placement of an order for replenishment of warehouse inventory and arrival of the goods at the warehouse receiving dock). This enables you to synthesize a realistic inventory versus time graph, exemplified by Exhibit C. Keep a running record of system status, using a tabulation sheet, graph, or preferably both.

Operate your model long enough to demonstrate to yourself that you know what you are doing. *Also*, identify questions you could answer by experiments on your model if time perimitted.

If time permitted, what are some of the "what if" questions that this model could answer?

Some "what if" questions that could be answered via this model

What is the consequence of lowering the reorder level?

Better yet, how much can the reorder level be lowered if the management will tolerate "stock outs" up to five days a year?

Of what consequence is an increase or decrease in the reorder quantity?

Will it help if the warehouse can find a source that responds more consistently?

Will it help if the warehouse can find a source (factory) that delivers faster?

Using a conversion table (based on Exhibit B) to generate daily sales, and a pie chart and spinner to generate replenishment times, an analyst produced the data shown in Exhibit D.

Exhibit D: Tab Sheet for Inventory Simulation

Day	1	2	3	4	5	6	7	8	9	10
Beginning-of-day inventory	280	269	253	245	230	222	210	193	336	325
Sales for the day	11	16	8	15	8	12	17	7	11	10
Receipts for the day	0	0	0	0	0	0	0	150	0	0
End-of-day inventory	269	253	245	230	222	210	193	336	325	315
Replenish?			YES		5 Days			←		

SELF QUIZ

1. List five uses to which engineers put models.

2. In what basic way does any model differ from its real-life counterpart?

3. List the three basic steps in developing a predictive model.

4. What is the primary limit on the amount of time that can generally be spent on refining a model?

5. List two differences between the way models are treated in engineering and the way they are treated by the sciences.

1. Engineers use models for thinking, communicating, predicting, controlling, and training.

2. Every model makes simplifying assumptions. Thus every model ignores some of the complexities of its real-life counterpart.

3. The three basic steps in developing a predictive model are

Construct, select, or modify a model.

Make observations of the real-life counterpart and compare with the model's predictions.

Interpret the observations and decide whether, with the discrepancies between the model and the real-life counterpart, the model should be applied as is, refined, or rejected.

4. Spending time to refine a model generally costs money and each succeeding round of refinement becomes more difficult, more expensive, and less productive, thus the relation of cost to benefits dictates the amount of refinement that should be done.

5. Engineers treat models as means to a solution for a problem. Scientists treat models as ends in themselves, or at least as things useful in themselves as explanatory or predictive concepts.

Engineers tend to cease refinement of models very quickly on benefit-cost ratio grounds. Scientists generally will continue refining models for long periods because the reason for developing the model has to do with longer-term scientific research purposes.

Learning Segment 9
OPTIMIZATION

LEARNING OBJECTIVES

- Understand the nature and purpose of optimization.
- Conduct an iterative optimization procedure in relation to typical technical and/or non-technical criteria.
- Be aware of the analytical optimization procedure and its advantages.
- Recognize the general applicability of the trade—off process, not only for engineering and related activities, but also for any situation involving conflicting aims or objectives.
- If your mathematical skills permit, use calculus to carry out analytical optimization procedures in relation to typical technical and/or non-technical criteria.

INTRODUCTION

When you adjust the focusing knob of binoculars in an attempt to sharpen the image, you are engaged in a process called optimization. Focus depends on the lens setting, which you manipulate until image sharpness is a maximum. In this and many other familiar cases, there is a *criterion* which is influenced by a *manipulated variable*, and a value of the manipulated variable for which the criterion is a maximum, called the *optimum value*. The manipulated variable in each of these cases has an optimum value with respect to the criterion indicated.

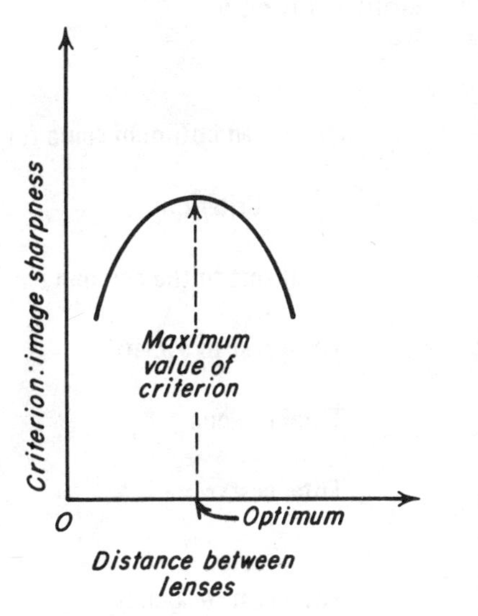

Manipulated Variable	Criterion
Fuel injection rate	Engine efficiency
Room temperature	Body comfort
Rate of work	Total work accomplished
Temperature	Bacteria multiplication rate

CHECK IT OUT

Cite 10 familiar situations in which there is obviously an optimum value for some variable with respect to a stated criterion. (For example, there is an optimum reading speed with respect to total knowledge assimilated.)

The answers to this exercise depend, of course, on your own experience, but some possibilities follow.

There is an optimum value for the independent variable in each of these situations.

Criterion	Manipulatable Variable
Total cost to the company	Number of maintenance men
Total cost to society	Level of air pollution
Total revenue	Product price
Total cost	Amount of protective packaging for a product during shipment
Total cost to society	Level of safety for a nuclear power plant

Optimum can be a minimum or it can be a maximum, so it should be viewed as *best* with respect to a given criterion, where best may mean maximization of the criterion (e.g., maximum engine efficiency) *or* it may mean minimization of the criterion (e.g., minimum with respect to cost). The concept of optimum is a basic one in engineering; there is an optimum solution to every problem. In fact, each specific characteristic of a solution has an optimum value. For example, there is an optimum size and shape for the coffee-pot handle with respect to ease of handling; an optimum process for refining petroleum with respect to total cost; and an optimum mix of ingredients with respect to the strength of concrete. Thus the concept of optimum permeates most aspects of an engineer's work. It guides actions and decisions, serving as a goal both for the solutions produced *and* for the way they are reached.

Optimization is the process of seeking the optimum solution. More specifically, it is an exploratory process involving a *search for* and *evaluation of* alternative solutions in order to locate the best one. Thus, in the search and decision phases of design, the engineer is engaged in optimization.

TECHNICAL TERMS

Write a brief definition for each term.

1. Criterion

____

____

2. Manipulatable variable

____

____

3. If you have not already done so, describe the relationship between criteria and manipulatable variables.

____

____

4. Optimization

____

____

1. A criterion is a basis of preference among alternatives. (It is a variable.)

2. A manipulatable variable (called solution variable in design jargon) is a solution characteristic that can be altered by the problem solver.

3. The relationship: the criterion depends on the manipulatable variable. (It is the latter that is altered to find the maximum or minimum value of the former.)

4. Optimization is a search for the best value of a manipulatable variable, i.e., for the one that maximizes (or minimizes) a given criterion. Optimization is the activity engaged in during the search and evaluation phases of the design process.

Unfortunately, in most engineering problems, optimization is more complex and time-consuming than in the binocular-focusing case. This is so especially because of numerous conflicting criteria. This problem is certainly not new to you. You have dealt with it if you have tried to tune in a television program when the clearest picture and clearest sound do not occur at the same setting of the dial. The optimum settings for picture clarity (O_p) and sound clarity (O_s) do not coincide; the setting that is optimum with respect to picture clarity is *suboptimum* for sound, and vice versa. In this situation, you must compromise between two conflicting criteria (picture clarity and sound clarity) conflicting in the sense that as you try to improve the situation with respect to one criterion, you are likely to make the situation worse with respect to the other. Before you can reach a satisfactory compromise, you must decide on the relative importance of picture clarity and sound clarity. Different persons will attach different degrees of importance to these two criteria, and so their compromise settings will differ, with most persons' settings falling somewhere between O_p and O_s.

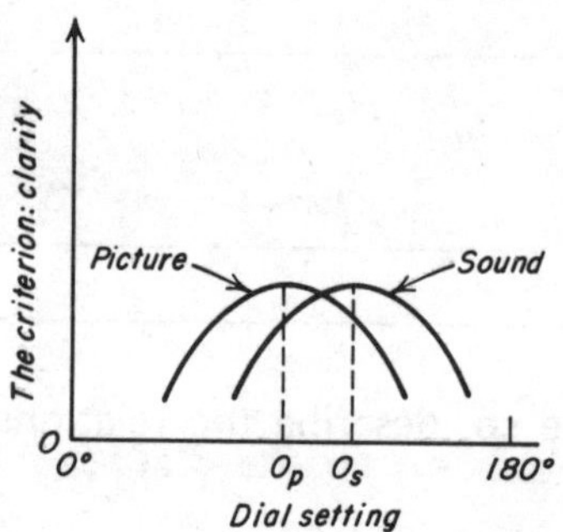

This situation (conflicting criteria and the need to find a compromise between them) abounds in engineering problems. Usually, however, the conflict is among many criteria. Take the engineer who is developing a machine for harvesting fruit. Some of the criteria he must consider are the speed with which the machine picks fruit, safety to persons near it, the degree to which it damages fruit, and cost. He cannot specify a machine that is the ultimate in speed (unless cost is no object and no one cares about safety and fruit damage), but the prospective purchasers of such machines *do* care. Hence he must sacrifice some speed to gain a reduction in cost, fruit damage, and hazards. Furthermore, he could design a machine that does not damage fruit, but it would probably be so slow and so expensive that no one would buy it. Similarly, the machine can be made perfectly safe, but only at a cost few potential buyers would consider paying; and so on for all criteria. Consequently, the engineer will alter his solution until he achieves what he believes is the optimum balance between these conflicting criteria.

This *is* a compromising process, and it gets complicated when there are more than two conflicting criteria or when they cannot be put in numerical terms. Yet it is a necessary process if the final solution is to be even close to optimum. In a relatively simple case it proceeds something like this. To determine the best balance between harvesting speed and damage to fruit, the engineer must know the relationship between these two criteria. On the basis of previous experience and some direct experimentation, this engineer estimates that damage depends on speed, approximately as shown here.

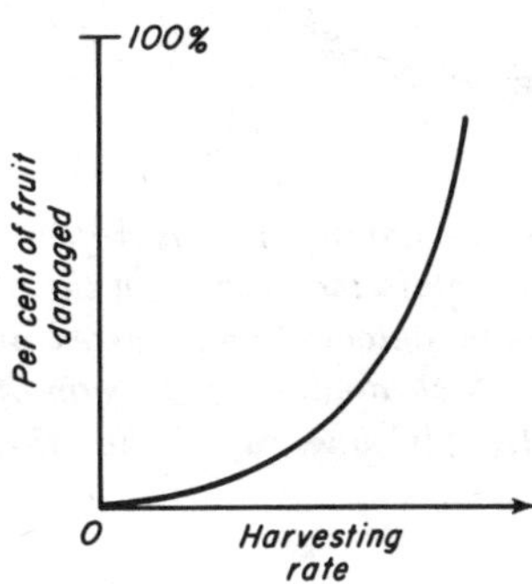

With the help of this *model* the engineer can predict the reduction in fruit damage that can be achieved by a given sacrifice in speed, or what a given increase in speed will cost in terms of damage. In the course of such deliberations, the engineer is determining how much speed to trade for reduced damage in order to achieve the optimum compromise. This "giving and taking" between conflicting criteria in order to reach the best balance is appropriately referred to as the *trade-off process*.

TRADE-OFFS

The trade-off process is all too familiar to the experienced engineer; almost all decisions involve trade-offs. The essence of this process is compromise, which is both essential and difficult. Some examples follow.

In the old days, an aircraft penetrating enemy territory was inclined to fly as high as possible to (hopefully) remain out of range. But today, with radar and antiaircraft missiles, the smartest tactic is to remain as low as possible. This has given rise to a new art for military pilots: terrain following–flying as low as possible, hugging the contour of the land. It also gives rise to the rather delicate trade-off situation graphed in Figure 34.

Recall the trestle of the bridge-tunnel described in Unit I. The choice of span between supporting piers illustrates the trade-off process and is also an interesting example of optimization in design. Figure 35 tells the story.

You can visualize some of the trade-offs that must be made in design of a mass transportation system. Criteria like convenience, speed, load-carrying capacity, and construction cost are conflicting and, in some instances, unquantifiable.

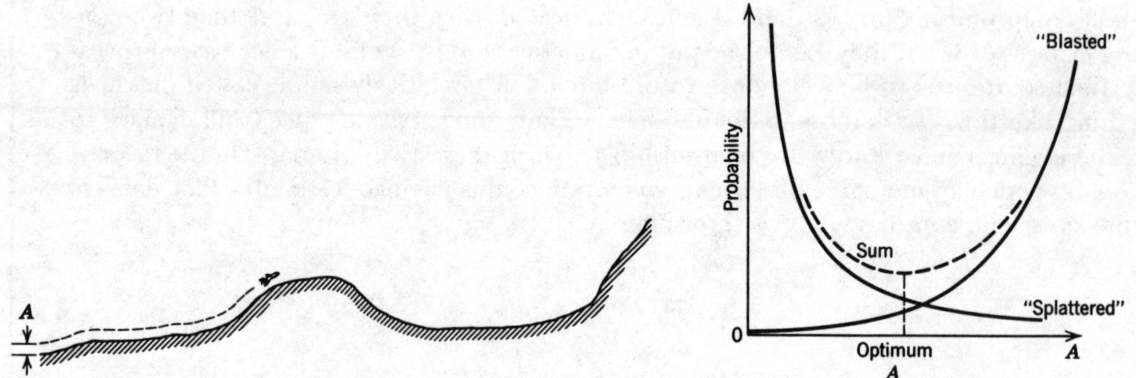

Figure 34. As the pilot flies closer to the ground (i.e., as A decreases), the probability of splattering himself on the side of a mountain increases. On the other hand, the higher he flies, the greater the probability of being detected and blasted out of the air with a missile. So he has a trade-off to make, which minimizes the sum of these two probabilities and, therefore, the probability of "getting it" one way or the other.

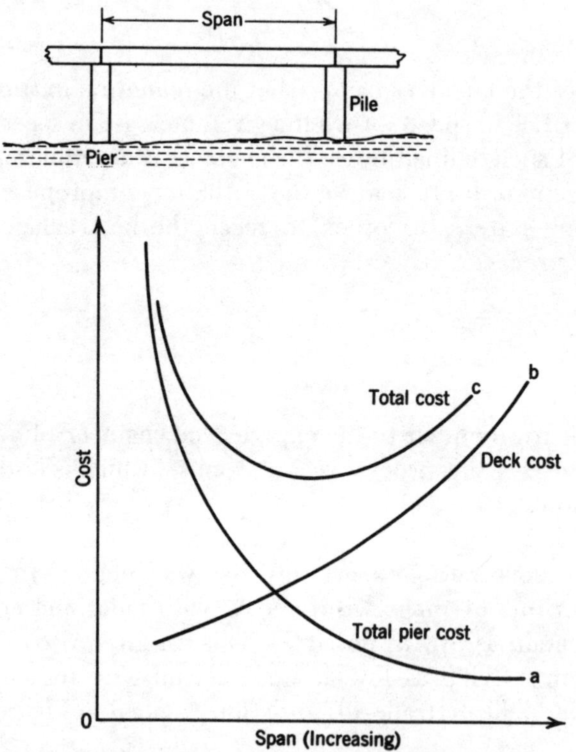

Figure 35. As indicated by curve a, if the distance between piers is increased, the total cost of piers diminishes, since fewer of them are needed for a given crossing. But there is a price to be paid as the span is lengthened: the roadway deck slabs must be sturdier, so their cost goes up, as indicated by curve b. It's another trade-off situation. The optimum deck length is determined by the minimum point on the total cost curve c.

YOU DECIDE

Trade-off situations are prevalent indeed. Think about the following circumstances and then record some of the trade-offs that must be made by the decision maker(s).

1. Energy supply versus environmental quality

____

____

____

2. Small college versus a large university

____

____

____

1. We cannot achieve the maximum in number of jobs, mechanical conveniences, and physical comfort and at the same time maximize air cleanliness and other measures of environmental quality. Compromises are necessary.

2. At a small college, you maximize criteria like student-faculty contact. At a large university, you maximize course selectivity, etc. When a student makes this choice, he has sacrificed in one respect or the other.

YOU DECIDE

For any two of the following, describe what you believe are important criteria that the designers must have considered. Identify criteria that appear to be conflicting and, therefore, between which trade-offs probably had to be made.

1. An interchange between two intersecting dual highways

2. An artificial hand

3. An automobile

Criteria for ________________ are ______________________________

____

____

____

and trade-offs must be made between ______________________________

____

____

____

Criteria for ____________ are ______________________________

____

____

____

and trade-offs must be made between ______________________________

____

____

____

You may recall that some of the criteria for these items were mentioned in an exercise in Unit II. Some of the criteria for the highway would include right-of-way cost, design and construction cost, social disruption, personal hardship, noise, pollution, aesthetic appeal, safety, maintenance cost, travel benefits provided, cost to users, and expandability. Trade-offs must be made between design-construction-land cost (i.e., first cost) and safety, first cost and expandability, first cost and aesthetic appeal.

Criteria for the design of an artificial hand include such items as cost to develop, cost to manufacture, functions performed and with what effectiveness, ease of operation, appearance, reliability, operating cost, skill required, and installation ease. Trade-offs must be made between first cost (development plus manufacturing costs) and variety of functions performed, first cost and effectiveness, first cost and reliability, first cost and ease of use, reliability and ease of use (increased reliability usually means more weight).

Criteria for the design of an automobile include development cost, manufacturing cost, visual appeal, efficiency, noise level, safety, maintenance cost, riding comfort, steerability, space, expected life, energy consumed, materials consumed, reliability, ease of operation, and disposability. Trade-offs must be made between first cost and performance, first cost and safety, first cost and durability, first cost and capacity, capacity and fuel consumption, durability (which when increased means a heavier vehicle) and fuel consumption, safety in major collisions and damageability in minor accidents, first cost and riding comfort.

VALUE DECISIONS

The engineer cannot arrive at the optimum trade-offs between criteria until he has estimates of the relative importance of each (recall the television tuning example). Knowing the relationship between harvesting speed and damage to fruit is of limited help until the engineer learns the relative importance of different speeds and degrees of damage to those who are potential users of this machine. Perhaps most users attach considerable value to high harvesting speed and are willing to tolerate a rather high percentage of damaged fruit in order to get this speed. Once the engineer knows the relative values attached to criteria, he is in a position to make trade-offs.

The assigning of relative value (importance, weight) to a criterion is commonly called a *value decision* (some say value judgment). Value decisions in engineering are difficult to make, partly because the engineer must anticipate the value attached to a given criterion *by others*, often a large group at that. For example, designers of a new-model automobile are contemplating the addition of a new safety feature. They must anticipate the relative values that potential buyers will attach to reduced risk of serious injury as opposed to X dollars this device will add to the price of the automobile.

Value decisions are never more difficult than when they involve human life. The designer of a highway must consider criteria such as construction cost and safety which, of course, are in conflict. He could specify a virtually impenetrable medial barrier for a four-land dual highway under design, thereby adding tremendously to total construction cost and reducing fatal accidents by two thirds. Is the added safety worth the cost? Ideally, tax payers should answer that question for the engineer, but it is hardly feasible for him to interview a large sample of taxpayers every time a decision like this is to be made. The only practical alternative for him is to second-guess the public's preference.

THINK IT THROUGH

The difficulty getting a sampling of taxpayer opinion is not the only difficulty facing the engineer trying to find a trade-off between dollars and safety in the preceding example. He is faced with a basic problem we have discussed before. What is it?

____

____

____

____

An answer to this exercise appears in the following paragraph.

This "dollars versus life" dilemma is only one of many instances in which criteria cannot feasibly be measured in common units. Dollars are convenient units, but many criteria are impractical to put in monetary terms—safety being a conspicuous example. Yet safety is a criterion to be reckoned with in most engineering problems. The inevitable result is that an engineer has some very trying value decisions to make.

OPTIMIZATION PROCEDURES

A traffic engineer is making a study with the objective of maximizing the number of vehicles that can pass through New York City's heavily traveled tunnels. (No, this is not simply a matter of speeding up the vehicles.) You know and so does he that drivers increase the spacing between vehicles as their speed increases. In fact, he has data (Figure 36) for a large sample of drivers indicating that spacing increases at a faster rate than speed, which leads him to suspect that encouraging drivers to move faster will increase throughput (vehicles per hour), but only up to a certain speed.

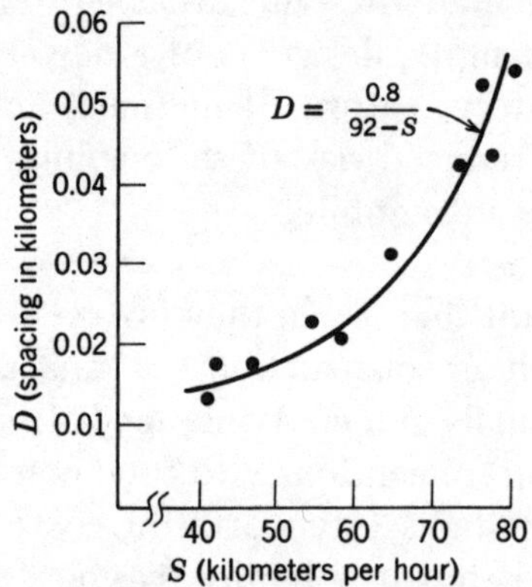

Figure 36. The engineer on this project obtained measurements of vehicle speed and spacing between successive vehicles from electronic apparatus set up in one of the City's tunnels. Each point on this graph represents a number of observations and can be interpreted as follows: for vehicles traveling around 70 kilometers/hour, the average spacing is 0.036 kilometers.

He expects that beyond that point, encouraging drivers to move faster will actually reduce the throughput because increased spacing cancels out the effect of increased speed. If his suspicion is correct, it follows that there is some average vehicle speed that is optimum (i.e., results in maximum throughput). He tested this hypothesis, proceeding in the following manner.

He developed the following mathematical model from his spacing versus speed data (Figure 36) by finding the equation describing the relationship between D (vehicle spacing) and S (speed) indicated by that data.

$$D = \frac{0.8}{92 - S} \quad \text{(a)}$$

THINK IT THROUGH

With formula (a), the engineer now has a mathematical model of the relationship between speed of vehicles and the average distance between successive vehicles. But the criterion (C) to be maximized is throughput, or rate of vehicles through the tunnel per hour per lane.

What formula involving this criterion (C), kilometers per hour (S), and kilometers occupied per vehicle (D) will give the relationship betwen the criterion and the manipulated variable?

The answer to this exercise appears in the following paragraph.

In order to convert this model into a form that describes the effect of S (the manipulated variable) on the criterion C (throughput in this case), he used

$$C = \frac{S}{D} = \frac{\text{kilometers per hour}}{\text{kilometers per vehicle}} \tag{b}$$

and substituted equation (a) into equation (b) thusly:

$$C = \frac{S}{D} = \frac{S}{\frac{0.8}{92 - S}} = 115\, S - 1.24\, S^2 \tag{c}$$

This mathematical model is called a *criterion function* (sometimes *objective function*). It describes in what way the criterion is a function of one or more manipulated variables.

THINK IT THROUGH

Can you anticipate what basic approach this engineer will take to finding the best value for speed to optimize the criterion of throughput?

An answer to this exercise appears in the following paragraph.

To determine what value of S maximizes C (i.e., what vehicle speed will get the most vehicles through the tunnel per interval of time), he substituted a trial series of speed values into equation (c) and computed the resulting values of C. The results are shown in Table A and Figure 37. S values in multiples of five were used to determine the general location of the maximum C, which appears to be near 45 kilometers per hour. Then, he explored a localized range of interest by substituting S values in increments of 1 kilometer per hour, enabling him to pinpoint the optimum S at 46 kilometers per hour (Table B, Figure 38).

Table A. Global Search

Trial Values of S in Kilometers per Hour	Resulting Values of C in Vehicles per Hour
35	2494
40	2600
45	2644
50	2625
55	2544

Figure 37. Global search, based on Table A.

Table B. Local Search

Trial Values of S in Kilometers per Hour	Resulting Values of C in Vehicles per Hour
41	2614
42	2625
43	2634
44	2640
45	2644
46	2645
47	2644
48	2640
49	2634

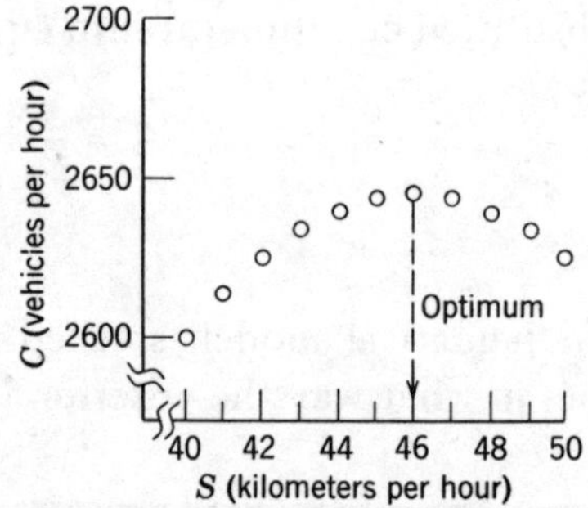

Figure 38. Local search, based on Table B.

The same conclusion can be reached by applying elementary calculus. Equation (c) is differentiated with respect to S, with this result:

$$\frac{dC}{dS} = 115 - 2(1.25)\,S = 115 - 2.5\,S \tag{d}$$

dC/dS is the rate at which C changes as S changes and, therefore, it is the slope of the curve in Figure 39. It is apparent that this slope is different at different values of S.

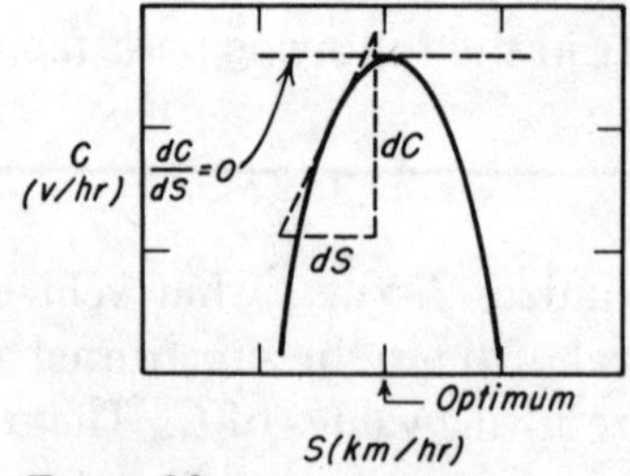

Figure 39.

At the peak of the curve, where C is a maximum, this slope is zero. Thus it remains to solve for the value of S at which

$$\frac{dC}{dS} = 0 \tag{e}$$

But

$$\frac{dC}{dS} = 115 - 2.5\,S$$

So

$$\begin{aligned} 115 - 2.5\,S &= 0 \\ S &= 46 \end{aligned} \tag{f}$$

Therefore, the optimum S is 46 kilometers per hour.

Iterative Optimization

The preceding two means of arriving at the optimum vehicle speed illustrate basically different methods of locating an optimum solution. One of these is the *iterative method* and it generally goes like this, using Figure 37 and 38 to illustrate.

I. The *global* search
   A. For the manipulated variable, the engineer assumes several values spaced over a range he suspects includes the optimum. (Five vehicle speeds over the 35 to 55 kilometers per-hour range were selected.)
   B. He predicts the effect of each of these assumed values on the criterion. [Equation (c) was used for this purpose, generating Figure 37.]

II. The *local* search
   A. Benefiting from the global search, more values of the manipulated variable are selected but over a narrower range. (Since Figure 37 indicated that the optimum is in the vicinity of 45 kilometers per hour, he chose values around it for investigation.)
   B. The effects on C are again predicted. [Using equation (c), he generated Figure 38.]

Ordinarily, this two-phase search will permit the engineer to estimate the optimum value for a manipulated variable satisfactorily. If not, the above procedure is repeated over a still narrower range.

The iterative method of optimization is basically an accelerated learning process. Through a series of successive approximations, the engineer gradually "closes in" on the optimum value of a manipulated variable, as diagrammed by Figure 40.

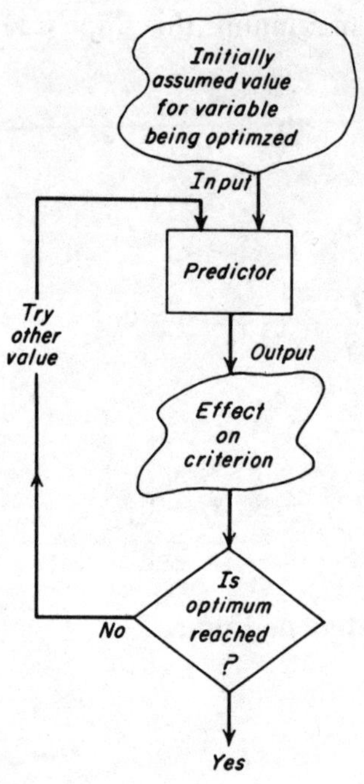

Figure 40.

LET'S BE SURE ABOUT THIS

Fill in each blank in the following outline with a brief description of the step in the iterative optimization method.

I. ______________________________

  A. ______________________________

  B. ______________________________

II. ______________________________

  A. ______________________________

  B. ______________________________

I. The global search
  A. Assume several values over the suspected range of the optimum.
  B. Predict the effect of each assumed value on the criterion.

II. The local search
  A. Assume several values over the narrower range revealed by the global search.
  B. Predict the effect of each value on the criterion and select the optimum value for the manipulated variable.

YOU DESIGN IT

Use an iterative optimization procedure to solve the following problem.

In laying telephone cable along the ocean floor, the chances of breakage by anchor snags and the like are reduced by allowing some slack in the cable. But cable costs money—plenty of it, in fact, so that you might think twice about "wasting" any cable. Being the perceptive person you are, you immediately identify this as an optimization problem. Determine the optimum amount of slack, assuming the following:

- D = linear miles of ocean traversed (Figure 41).
- L = length of cable used.
- S = slack = L-D.

That a new cable installation is under design, where $D = 80$ kilometers, for which cable cost is \$4000 per kilometer and for which the breakage cost (i.e., cost of interrupted service and repairs due to breaks over the life of the cable) depends on slack; this function is estimated to be

$$\text{Breakage cost} = \frac{\$1{,}000{,}000}{S^2}$$

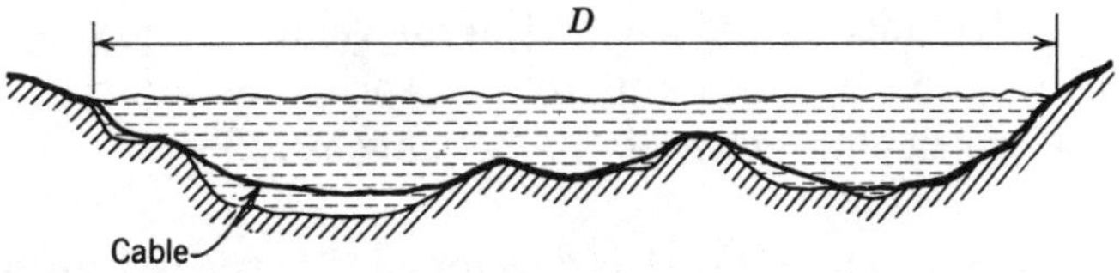

Figure 41.

Plot *breakage cost* as a function of slack, *cable cost* as a function of slack, and then *total cost*. Be careful in your choice of scales.

	S	Breakage cost = $\frac{1{,}000{,}000}{S^2}$	Extra cable cost = 4,000 (S)	Total cost
Global				
Local				

One possible solution.

$$\text{Total cost} = (\text{breakage cost}) + (\text{extra cable cost}) = \frac{\$1{,}000{,}000}{S^2} + \$4{,}000\ (S)$$

S	Breakage cost $\frac{1{,}000{,}000}{S^2}$	Extra cable cost 4,000 (S)	Total cost
5	$40,000	$20,000	$60,000
10	10,000	40,000	50,000
15	4,444	60,000	64,444
6	27,800	24,000	51,800
7	20,400	28,000	48,400
8	15,600	32,000	47,600
9	12,350	36,000	48,350

Analytical Optimization

The other of the two basic procedures for locating an optimum is the *analytical method*. In this case, a mathematical model yields the optimum value directly, as illustrated by the second method of arriving at the optimum vehicle speed in tunnels. Often the model is derived through the use of calculus, generally proceeding as follows.

1. A criterion function [e.g., equation (c)] is developed. Its general form is

$$C = f(V)$$

where C is the criterion and V is the manipulated variable. Although C is often measured in dollars, it need not be. It could be number of passengers carried, pounds, or efficiency.

2. Then, by differential calculus or other means, the criterion function is converted to a form that yields the optimum value of the manipulated variable directly.

The criterion function can contain more than one variable to be optimized, in either the iterative or the analytical method. The general form for the criterion function, then, is

$$C = f(V_1, V_2, V_3, \ldots)$$

where V is a manipulated variable.

Surely you wonder why anyone would use the iterative procedure, considering the directness of the analytical method. It's a good question and there is a good reason: in many cases, the analytical method is much too difficult mathematically and the iterative method is the only practical way. Furthermore, in many instances the engineer wishes to know how *sensitive* the creation is to deviations of a variable from its optimum value, and the iterative method ordinarily provides this information conveniently.

THINK IT THROUGH

Can you think of any reasons why an engineer would want to know how sensitive the criterion is to deviations of a manipulated variable from the optimum?

____

____

____

One basic answer for exercise appears in the following paragraph.

Note from the preceding Table A that very little is lost in terms of vehicles per hour if the average speed should deviate slightly from the 46 kilometers per hour optimum. If the average speed is 50 kilometers per hour, there is a loss of only 20 vehicles per hour, which is a drop of less than 1 percent. This information is useful because it would be difficult to attempt to control the average speed to 46 kilometers per hour and so the engineers were interested in learning what sacrifice in vehicles per hour would be made if they attempted to control speed to a higher value. An investigation of this kind, to learn the consequences of setting a variable at something other than its optimum value, is a *sensitivity analysis*.

THINK IT THROUGH

With the term sensitivity analysis in mind, what does this graph indicate to you?

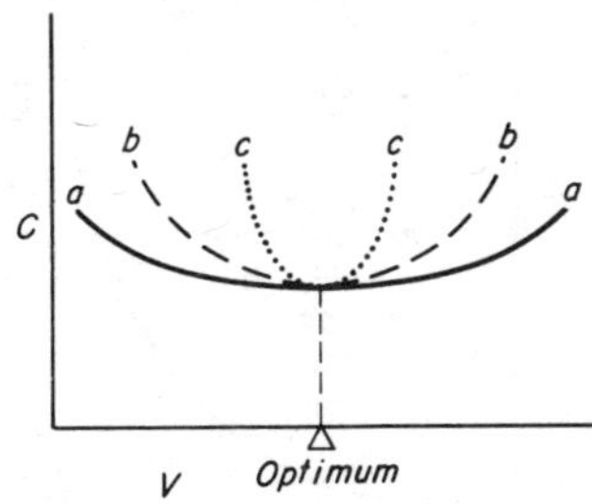

This graph shows three different criteria, (a), (b), and (c), which depend on a manipulatable variable. The criterion (c) is quite sensitive to changes in the manipulatable variable. In contrast, criterion (a) is relatively insensitive; the manipulatable variable can vary over a rather wide range with very little effect on C.

The formal optimization methods cannot be relied on as much as the engineer would like. Because of a preponderance of unquantifiable criteria, the large number of variables, or a lack of time, engineers often must rely on procedures that are less formal, less quantitative, and less objective than those illustrated. Usually they use a variety of different methods in their optimizing efforts, the particular combination of procedures varying considerably from problem to problem (so much so that it is foolish for an author to attempt to generalize any more than we have).

Optimization of Problem-Solving Methods

The concept of optimum, applied up to now only to solutions to problems, is just as applicable to the *methods* the engineer employs in arriving at those solutions (e.g., the measurement systems, the computational methods, the models, and the number and types of technicians utilized).

Over the long run, the errors in a model's predictions cost something. This is so because of mistakes, failures, accidents, repairs, and alterations that result when decisions are based on erratic predictions *or* because of the high safety factors necessary to protect against such adverse occurrences. And yet it is worth reducing these errors through model refinement only up to a point. For instance, the designers of a chemical plant are relying on a model to make predictions on which to base their design. If, after the plant has been built, there is a small disagreement between performance predicted and performance experienced, it is of little practical consequence and is considered inevitable. In this instance the cost of the lack of correlation between predicted and actual results is negligible, as indicated by point 1 on curve *a* of Figure 42. A larger discrepancy, however, might well result in some wrong decisions that are discovered only after the plant is built, the penalty for which is some costly alterations. In this case, the cost of the lack of correlation may be somewhere around point 2 of curve *a*. A still larger lack of correlation could result in something much more costly, an explosion, the cost of which might be in the area of point 3. Curve *a* indicates what generally happens to the cost of errors in a model's predictions as the engineer refines the model and reduces those errors. The cost declines, but at a decreasing rate, and so a point is eventually reached at which additional refinements are of negiligible benefit and not worth striving for.

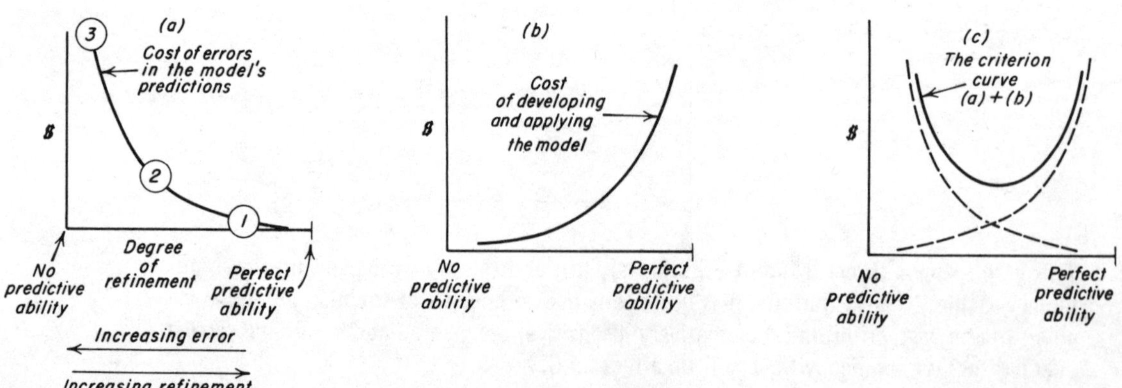

Figure 42.

There is another good reason for not attempting to refine a model to the point where curve *a* levels off: the cost of developing and applying it increases, as indicated by curve *b* of Figure 42. As additional efforts are expended to reduce the error in a model's predictions, this cost accelerates upward because additional improvements become more and more difficult and time-consuming to achieve. This is generally true.

The optimum degree of refinement, then, is the point at which the sum of these two costs is a minimum, as shown by curve *c*. It is uneconomical to attempt to refine this model beyond that point; likewise for all models. Therefore, there is an optimum degree to which a model should be refined.

The situation summarized by curve *c* is not unique to modeling. The same holds true for measurement systems, information–search methods, and most tools and techniques. In each case, there is an optimum degree of refinement simply because the manpower and other resources devoted to such efforts are valuable for other purposes and because continued refinement efforts bring diminishing returns. For these reasons, there is an optimum number of man-hours to devote to the solving of a problem.

LET'S BE SURE ABOUT THIS

What are the two reasons (represented by curve *a* and curve *b* in the preceding graphs) for limiting the degree to which a model is refined?

The two reasons for limiting the degree to which a model is refined have to do with the cost of refining the model and the cost of errors in the model. As the model becomes increasingly refined, the cost for each unit of improvement (marginal cost) becomes greater to the point where more is spent for the improvement than the improvement could possibly be worth. Also, as the model becomes more and more accurate as a representation of its real-life counterpart, the cost resulting from any error in its representational ability becomes less and less until no cost due to the error could exceed the cost to improve the model.

Optimum As a Goal

An important goal in engineering endeavors is that of the "optimum." Engineers strive to produce optimum solutions to problems and to do so by optimum means. Notice the word *strive*. Although the optimum solution is almost always an objective, it is seldom a realization. Many real-world problems are too complex for an optimum solution to be located in a reasonable period of time. In many instances, the amount of time required would be longer than the life of the problem. Invariably, many other problems are awaiting attention, so that it often becomes more profitable to turn to one of these than to continue searching until optimum is achieved for the current problem. Thus, in design, it is usually a matter of progressing toward the optimum, continually seeking better solutions until the effort becomes more profitably spent elsewhere.

APPLICABILITY OF OPTIMIZATION CONCEPTS AND TECHNIQUES

You can capitalize on this brief introduction to optimization simply by calling on your knowledge of algebra, graphs, and other relatively elementary aspects of modeling; you can do even more with some knowledge of calculus. These concepts have a surprisingly broad range of applications, which includes a variety of nontechnical problems. Here is an illustration of how you can apply the contents of this unit to a problem you do not associate with engineering.

CASE STUDY: NEVERSMUDGE PRINTING COMPANY

Periodically, the Neversmudge Printing Company must replenish its supply of paper. The frequency with which it does so and the quantity of paper it buys at one time significantly affect the cost of their paper. For instance, each time they replenish their paper stock, certain costs are incurred due to paperwork, handling, and other activities. The total of these costs, the *acquisition cost*, is relatively constant, regardless of the amount of paper ordered. This is $25 per purchase in Neversmudge's case.

Of course, it costs money to store the paper, since it has to be housed and insured, and because of interest on the money tied up in inventories. The total of these costs is called the *storage cost*. In Neversmudge's case, this amounts to $4 per year for a roll of paper.

If the company orders one roll at a time (as needed) they can keep the total storage cost to a bare minimum but, in doing so, they will inflate the total acquisition cost ridiculously. Or they can order a whole year's supply of paper at once, which minimizes the acquisition cost but results in a prohibitively high yearly storage cost. Somewhere between extremes is a "compromise purchase quantity," which results in minimum *total* cost to the company. Knowing something about optimization, either calculus and or graphs, *you* could find the optimum purchase quantity for them, by the following procedure.

First, you know that the total annual cost in this case is

$$T = \text{(yearly acquisition cost)} + \text{(yearly storage cost)*} \tag{g}$$

$$= \begin{pmatrix}\text{number of orders}\\ \text{placed in a year}\end{pmatrix} \begin{pmatrix}\text{acquisition}\\ \text{cost}\end{pmatrix} + \begin{pmatrix}\text{number of rolls}\\ \text{ordered per}\\ \text{purchase}\end{pmatrix} \begin{pmatrix}\text{storage}\\ \text{cost}\end{pmatrix}$$

$$= \begin{pmatrix}\dfrac{\begin{matrix}\text{rolls of paper}\\ \text{needed per}\\ \text{year}\end{matrix}}{\begin{matrix}\text{number of rolls}\\ \text{ordered per}\\ \text{purchase}\end{matrix}}\end{pmatrix} \begin{pmatrix}\text{acquisition}\\ \text{cost}\end{pmatrix} + \begin{pmatrix}\text{number of rolls}\\ \text{ordered per}\\ \text{purchase}\end{pmatrix} \begin{pmatrix}\text{storage}\\ \text{cost}\end{pmatrix}$$

Since algebraic manipulations in terms of words are awkward to say the least, surely you would convert the above expressions to symbols like these before proceeding.

- N = number of rolls of paper required per year (3600 in this instance).
- A = acquisition cost per order ($25 in this case).
- S = storage cost, in dollars per roll per year ($4 in this instance).
- Q = the number of rolls per purchase, called the purchase quantity (which you are seeking to optimize).

*In this example, the cost of the material itself is ignored, since it is the same for all purchase quantities and therefore irrelevant to this decision.

This enables you to express equation (g), the criterion function, as

$$T = \left(\frac{N}{Q}\right) A + (S \times Q) \tag{h}$$

This case illustrates the types of problems that *you* are equipped to handle, with or without a knowledge of calculus. You can prove this to yourself by completing the following exercise.

YOU SOLVE IT

This problem can be solved with the methods of differential calculus, of course. We will examine the general approach to a solution via differentiation later. But now, use iterative methods to solve this problem for Neversmudge's purchasing manager. (A solution to the closest multiple of 50 rolls is sufficient.)

Q Number of rolls per purchase	$\frac{N}{Q}$ Number of purchases per year	$\frac{N}{Q}A$ Yearly acquisition cost	$S(Q)$ Storage cost per year	$T = \frac{N}{Q}A + S(Q)$ Total yearly cost
600	$\frac{3600}{600} = 6$	6 x \$25 = \$150	600 x \$4 = \$2400	\$150 x \$2400 = \$2550

Determination of Optimum Purchase Quantity by the Iterative Method

Q Number of rolls per purchase	$\frac{N}{Q}$ Number of purchases per year	$\frac{N}{Q}A$ Yearly acquisition cost	$S(Q)$ Storage cost per year	$T = \frac{N}{Q}A + S(Q)$ Total yearly cost
600	$\frac{3600}{600} = 6$	6 X \$25 = \$150	600 X \$4 = \$2400	\$150 + \$2400 = \$2550
200	18	450	800	1250
150	24	600	600	1200
100	36	900	400	1300
50	72	1800	200	2000
(a)	(b)	(c)	(d)	(e)

A graphic view of columns c, d, and e.

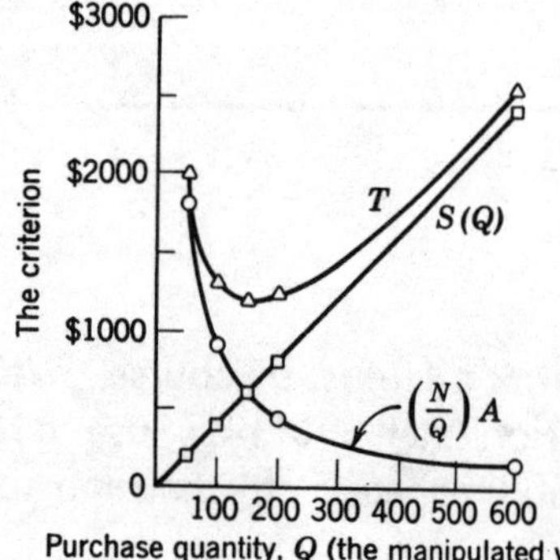

Bravo! In a few minutes of pencil-and-paper work, you have the answer for the manager of purchases. He is impressed, but he is also skeptical, since he prefers the analytical methods of calculus. But you, if you know calculus, can get around that; you prepare the following equations.

You may also know that equation (h) can be differentiated with respect to Q.

$$\frac{dT}{dQ} = \frac{-N\,(A)}{Q^2} + S \tag{i}$$

and that you can find the value of Q for which T is a minimum by the same method outlined earlier.* And so:

$$\frac{dT}{dQ} = \frac{-N\,(A)}{Q^2} + S = 0 \tag{j}$$

$$Q = \sqrt{\frac{N \times A}{S}} \tag{k}$$

Thus, you have found a simple algebraic expression, equation (k), that yields the *optimum* purchase quantity (i.e., the purchase quantity that minimizes total cost). By substituting numerical values, you calculate the optimum Q value as

$$\sqrt{\frac{3600\,(25)}{4}} \text{ or 150 rolls}$$

*When $dT/dQ = 0$, T is either at a maximum or a minimum. In this instance you are familiar with the function and happen to know that equation (j) represents a minimum value for T. However, in the absence of such knowledge, you could test for maximum or minimum by taking the second derivative of equation (i) and determining whether it is positive or negative at the point in question.

Optimum, trade-off, and optimization should come to mind when you think and talk about some of the major techno-social problems of the times. Certainly there are major trade-offs to be made in solving environment and energy problems, trade-offs that are occasionally overlooked in the demands and debates associated with these issues. Can persons who speak of "zero air pollution" be conscious of the trade-off that must be made between cost of abatement and cost of polluted air? True, they are not always made explicit, even recognized, but trade-offs pervade political, personal, business, military, medical, and all other forms of decision making. So your encounter with trade-offs and other aspects of optimization is hardly going to terminate with the completion of this, program.

SELF QUIZ

1. Define the term *optimization*.

2. Define the term *manipulatable variable*.

3. Define the term *criterion* as used in the optimization process.

4. Briefly describe what the *trade-off process* is to an engineer.

5. Briefly describe the *iterative optimization procedure*.

6. In just a few words, what is *analytical optimization*?

1. Optimization is a search for the optimum (best) value of a manipulatable variable, i.e., for the one that maximizes (or minimizes) a given criterion. Optimization is the activity engaged in during the search and evaluation phases of the design process.

2. A manipulatable variable (called solution variable in design jargon) is a solution characteristic that can be altered by the problem solver.

3. A criterion is a basis of preference among alternatives. (It is a variable.)

4. The trade-off process is necessary for the engineer whenever there are conflicting objectives to be met by a solution. Then the degree of attainment of one objective must be balanced (traded off) against the degree of attainment of the conflicting objective(s).

5. The iterative optimization procedure may be briefly described as follows.

I. The global search
   - A. Assume several values over the suspected range of the optimum for the manipulatable variable.
   - B. Predict the effect of each assumed value on the criterion.

II. The local search
   - A. Assume several values for the manipulatable variable over the narrower range revealed by the global search.
   - B. Predict the effect of each of these values on the criterion until the optimum value (within an acceptable range) is found for the variable in relation to the optimum for the criterion.

6. Analytical optimization is the use of calculus to compute the optimum directly.

ISBN 0-471-01702-7